Nature's Blueprints: Unraveling the Secrets of Crystal Design

Manasa

Contents

1. Introduction..1

1.1 B20 compounds and magnetic skyrmions 1

1.2 B20 MnSi with magnetic Skyrmions..8

1.2.1 Crystallization process 10

1.2.2 Phase diagram of Mn-Si binary compounds..13

1.2.3 Bulk B20-MnSi..14

1.2.4 B20-MnSi thin film..18

1.2.5 B20-MnSi nanowire..25

1.3 Fast annealing method 27

1.4 Objectives and the structure of the book 30

2. Experiment 32

2.1 Sample preparation 32

2.1.1 DC magnetron sputtering..32

2.1.2 Sub-second annealing 35

2.2 Structure characterization: X-ray diffraction..40

2.3 Property characterization 41

2.3.1 Magnetic properties 41

2.3.2 Magneto-transport properties..44

3. B20-MnSi films grown on Si(100) substrates with magnetic skyrmion signature.......46

3.1 Introduction...46

3.2 Experiment..47

3.3 Results and Discussions...48

3.4 Conclusions...56

4. Phase selection in Mn-Si alloys by fast solid-state reaction with enhanced skyrmion stability..**57**

4.1 Introduction..57

4.2 Experiment..59

4.3 Results..61

4.3.1 MnSi and MnSi$_{1.7}$ phase reaction..61

4.3.2 Magnetic Skyrmion..68

4.3.3 Discussion..76

4.4 Conclusion..78

5. On the Curie temperature of MnSi films..**80**

5.1 Introduction..80

5.2 Experiment..82

5.3 Results..83

5.4 Conclusion..89

6. Summary and outlook **90**

6.1 Summary..90

6.2 Outlook 91

6.2.1 Film thickness effect on formation of (111)-textured B20-MnSi..........................91

6.2.2 MnSi$_{1.7}$% influence on Skyrmion stability 96

6.2.3 Preparation of other transition-metal monosilicides and germanides....................98

Acknowledgement..**99**

References..**101**

1. Introduction

1.1 B20 compounds and magnetic skyrmions

Crystal structure is the description of an ordered arrangement of atoms, ions or molecules in a crystalline material [1] shown schematically in Figure 1.1 (a). Ordered structures occur from the intrinsic nature of the constituent particles to form symmetric patterns. Those patterns repeat along the principal directions of three-dimensional space in materials. The smallest group of particles in the material that constitutes this repeating pattern is the unit cell (in Figure1.1 (b)) of the structure. The unit cell completely reflects the symmetry and structure of the entire crystal, which is built up by repetitive translation of the unit cell along its principal axes. The lengths of the principal axes (a, b and c) of the unit cell and the angles (α, β and γ) between them are the lattice constants, also called lattice parameters. A lattice system is a class of lattices with the same set of lattice point groups, which are subgroups of the arithmetic crystal classes. There are seven lattice systems (triclinic, monoclinic, orthorhombic, tetragonal, rhombohedral, hexagonal, and cubic) classified by lattice constants [2]. They are listed in Table 1.1. Depending on the symmetry properties of the crystal, the seven lattice systems are divided into space groups. All possible symmetric arrangements of particles in three-dimensional space are described by 230 space groups.

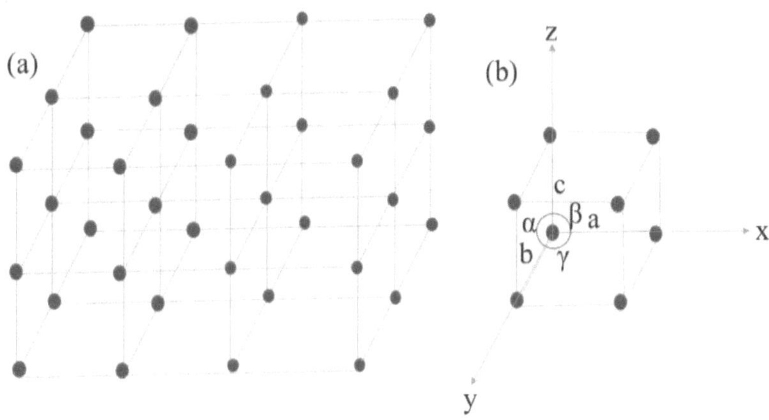

Figure 1.1. Crystal structure and unit cell.

Table 1.1. Seven lattice systems

Crystal system	Length of edges	angles
Triclinic	$a \neq b \neq c$	$\alpha \neq \beta \neq \gamma \neq 90°$
Monoclinic	$a \neq b \neq c$	$\alpha = \gamma = 90°, \beta \neq 90°$
Orthorhombic	$a \neq b \neq c$	$\alpha = \beta = \gamma = 90°$
Tetragonal	$a = b \neq c$	$\alpha = \beta = \gamma = 90°$
Trigonal	$a = b \neq c$	$\alpha = \beta = 90°, \gamma = 120°$
Hexagonal	$a = b \neq c$	$\alpha = \beta = 90°, \gamma = 120°$
Cubic	$a = b = c$	$\alpha = \beta = \gamma = 90°$

The B20-structure is one kind of simple cubic structure and belongs to space group P213, #198 [3, 4]. It is a distorted rock salt structure with basis vectors (u, u, u), $(\frac{1}{2} + u, \frac{1}{2} - u, -u)$, $(-u, \frac{1}{2} + u, \frac{1}{2} - u)$ and $(\frac{1}{2} - u, -u, \frac{1}{2} + u)$. As shown in Figure 1.2, 4 transition metal atoms and 4 IV group atoms constitute the unit cell of a B20 crystal. Compared with the regular cubic structure, the nearest neighbouring atom of any atom in B20 compounds is a different element from itself. Due to this special arrangement of these atoms, the spatial inversion symmetry is broken and thus B20 compounds are noncentrosymmetric or called chiral symmetrical [5]. This broken symmetry is a fundamental concept prevailing in many branches of physics. In recent years, chiral B20 compounds are found to be carrier for skyrmions [6-9] and Weyl fermions [10-14].

B20 crystal structure

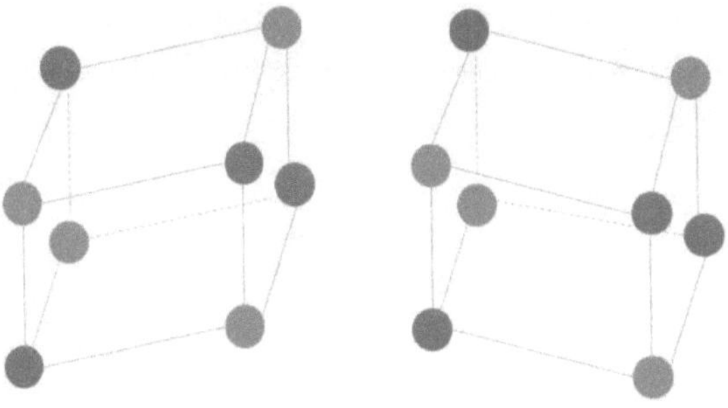

Figure 1.2. B20 crystal structure. The nearest atom of each atom is different. Thus this special atom arrangement breaks the spatial inversion symmetry, also called noncentrosymmetry. The left and right unit cell indicate the existence of chirality along one plane.

Due to the existence of antisymmetric exchange interaction, B20 compounds have many interesting magnetic, or transport properties originating from the relativistic spin orbit coupling (SOC) [15-18].

The interplay between SOC and magnetism can lead to many interesting phenomena. In conventional magnetic materials, the ferromagnetic order arising from exchange interactions aligns neighboring spins [15]. A well-known result of SOC is magnetocrystalline anisotropy. The electron moments tend to align preferentially along certain crystallographic directions (called easy axes) due to the coupling of electron motion to the crystalline lattice field. In systems, for instance B20 compounds that do not exhibit inversion symmetry, SOC induces a chiral Dzyaloshinskii-Moriya interaction (DMI) [8, 9, 19-21]. DMI contributes additionally to the total magnetic exchange interaction between two neighboring magnetic spins, S_i and S_j. Figure 1.3 (a) shows the schematic of DMI. The Hamiltonian can be written as [22-24]:

$$H_{DM} = D_{ij} \cdot (S_i \times S_j) \tag{1}$$

Here S_i and S_j are neighbouring spins and D_{ij} is the Dzyaloshinskii–Moriya vector. If S_i and S_j are initially parallel, the effect of a sufficiently strong DMI (with respect to exchange and anisotropy) is to introduce a tilt around D_{ij}. DMI can enable the formation of chiral spin structures, for instance, a non-collinear structure shown in Figure 1.3 (b) and magnetic skyrmions [15, 16, 23, 25, 26].

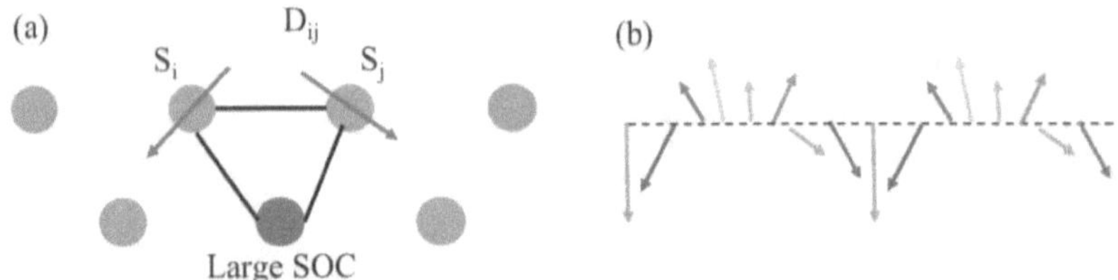

Figure 1.3 (a) Dzyaloshinskii Moriya interaction and (b) induced non-collinear spins. Reproduced from [15].

Skyrmion is a topologically stable field configuration of a certain class of non-collinear sigma models, which was proposed by Tony Skyrme in 1962 [27]. Figure 1.4 (a) shows the three-dimensional spin arrangement of a magnetic skyrmion, which looks like a hedgehog ball . Figure 1.4 (b), (c) and (d) are the two-dimensional schematic diagram of the magnetic skyrmion [28-30]. The spins of magnetic skyrmion point down at the center and up at the edges. There are three typical types of magnetic skyrmion: Bloch-type skyrmion (Figure 1.4 (b)), Néel-type skyrmion (Figure 1.4 (c)) and Anti-skyrmion (Figure 1.4 (d)). They correspond to different symmetries of the interaction between spins, resulting in different directions of the rotation. The interaction between spins can be due to the underlying crystal lattice or to the presence of an interface. The topological number or the skyrmion number can be defined as [31-33]:

$$N_{sk} = \frac{1}{4\pi} \int n \cdot \left[\frac{\partial n}{\partial x} \times \frac{\partial n}{\partial y} \right] dxdy = \pm 1 \tag{2}$$

where n is an unit vector parallel to the local magnetic moment.

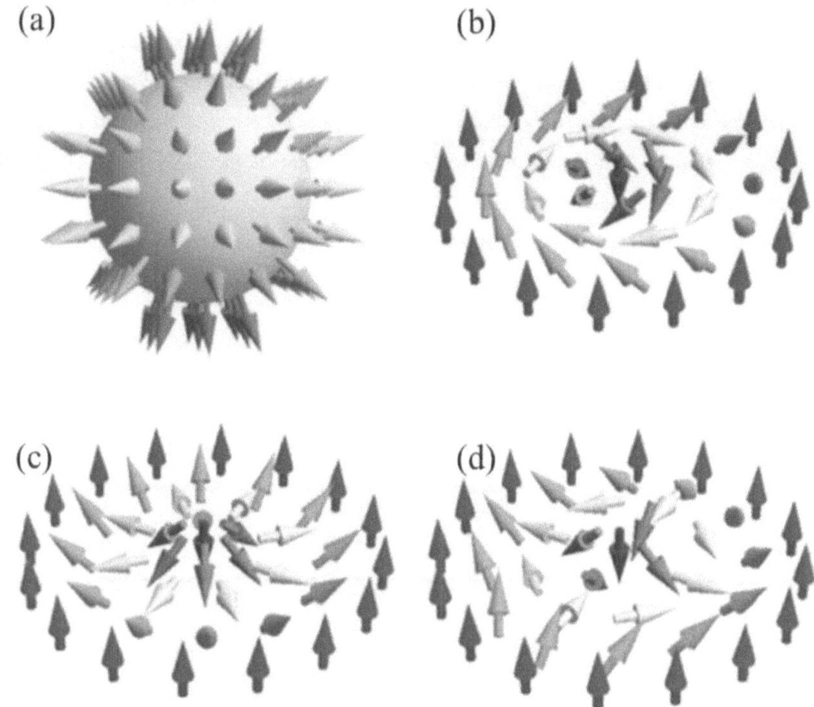

Figure 1.4. (a) Hedgehog-type skyrmion spin texture and (b–d) its projection into the two-dimensional plane (Bloch-type Skyrmion (b), Néel-type Skyrmion (c) and Anti-Skyrmion (d)). Depending on the underlying lattice symmetry (i.e. crystallographic point group), Dzyaloshinskii-Moriya interaction can stabilize different forms of skyrmion spin texture. Reproduced from [28].

The topological number for Bloch-type skyrmion and Néel-type skyrmion is 1, while for the Anti-skyrmion it is -1 [28, 34-38]. The mechanism for forming magnetic skyrmion can be classified into 4 kinds: (1) the interactions between magnetic dipoles in centrosymmetric magnets. If there is a perpendicular magnetic anisotropy in magnetic films [18, 25, 32, 39-41]. The magnetocrystalline anisotropy causes the magnetic spins to be arranged perpendicular to the surface of film and the interactions between dipoles make spins align the in the plane preferably. When the applied magnetic field is perpendicular to the surface of film, the stripe domains become periodically arranged magnetic skyrmions; (2) DM interactions in noncentrosymmetric materials, like, MnSi, MnGe, $Fe_{1-x}Co_xSi$, FeGe and Cu_2OSeO_3 [5, 7, 8,

42, 43]. Under the external field, the magnetic domains of these materials change from spiral to triangular skyrmion; (3) frustrated exchange interactions [34]; (4) exchange interactions between 4 spins [22]. In the materials with the first mechanism, the competition between dipole interaction and the exchange interaction determine the size of skyrmion, between 100 nm and 1 μm. In materials with the second mechanism, the ratio of the DM interaction and the exchange interaction determine the size of skyrmion, between 2 nm and 100 nm. In the materials with the third or fourth mechanisms, the size of skyrmion can be on an atomic scale.

Up to now, the materials for hosting skyrmion can be classified into 3 kinds: noncentrosymmetric magnetic, centrosymmetric magnetic and interface magnetic. The parameters of these three kind materials are shown in table 1.2.

Table 1.2. The list of different materials that host magnetic skyrmions.

Classification	Materials	Transition temperatures	Size of skyrmion	Transport properties	Ref
Noncentrosymmetric Magnetic	MnSi	29.5 K	18 nm	Metal	[5]
	FeCoSi	<36 K	100 nm	Semiconductor	[42]
	FeGe	278 K	70 nm	Metal	[44]
	MnGe	170 K	3 nm	Metal	[45]
	$Co_xZn_yMn_{20-x-y}$	>300 K	>120 nm	Metal	[46]
	$Mn_{1.4}Pt_xPd_{1-x}Sn$	>300 K	150 nm	metal	[47]
	Cu_2OSeO_3	59 K	62 nm	Insulator	[48]
Centrosymmetric Magnetic	$Ba(FeScMg)_{12}O_{19}$	>300 K	200 nm	Insulator	[49]
	LaSrMnO	100 K	160 nm	Insulator	[50]
	MnNiGa	350 K	180 nm	Metal	[51]
Interface magnetic	Fe/Ir (111) film	>300 K	1 nm	Metal	[22]
	Co/Ni/Cu (001)	300 K	2 μm	Metal	[33]
	[Pt/Co]/Pt	300 K	120 nm	Metal	[52]

Here, we make a brief introduction about the formation mechanism of nanometric skyrmions in B20-type magnetic compounds, that have been reported experimentally. Based on the previous discussion, in B20-type magnets, there exists an antisymmetric exchange interaction [50, 53, 54], i.e. Dzyaloshinskii Moriya (DM) interaction originating from relativistic spin orbit interaction. From Eq. 1, D_{ij} is the DMI contribution, also called as the DM factor. For most of magnetic materials, their spins tend to arrange in a collinear fashion due to Heisenberg exchange interaction [25, 55-57]. Dzyaloshinskii Moriya interactions and Heisenberg exchange interaction compete with each other, forming a whirling spin structure. The size of skyrmions or the associated magnetic modulation period λ is determined by the ratio between the magnitudes of DM interaction D_{ij} and ferromagnetic exchange interaction J_{ij}. Ferromagnetic exchange interaction is an interaction between localized electron magnetic moments, S_i and S_j. Its Hamiltonian can be written as [15]:

$$H_{Fe} = J_{ij} S_i \cdot S_j \tag{3}$$

Here S_i and S_j are neighbouring spins from neighbouring atoms and J_{ij} is the ferromagnetic exchange integral. The skyrmion size is determined by the ratio between Heisenberg and DM interaction [15]:

$$\lambda \sim a \cdot \frac{J_{ij}}{D_{ij}} \tag{4}$$

where a is the lattice constant. For ferromagnetic B20 compounds like MnSi, FeGe [58, 59], $J_{ij} \gg D_{ij}$ and for antiferromagnetic B20 compounds like MnGe, $J_{ij} \ll D_{ij}$, thus MnGe has a much smaller skyrmion size close to 2 nm and MnSi or FeGe have much bigger skyrmions larger than 10 nm. The size of skyrmion also varies slightly in the bulk, thin film and nanowires samples due to the change of lattice constant and exchange interactions. λ typically ranges from 1 to 100 nm for noncentrosymmetric compounds. The saturation field H_s for these materials is also can be estimated as [15]:

$$H_s \sim \frac{D_{ij}^2 M}{J_{ij}} \tag{5}$$

where M is the saturation magnetisation. For ferromagnetic B20 compounds, like MnSi, FeGe the $J_{ij} \gg D_{ij}$. Thus the saturation field H_s is always smaller than 1 T. However, for antiferromagnetic materials, MnGe [45] H_s is more than 14 T due to smaller J_{ij}. From these equations, we can calculate and design the size of skyrmions and the saturation field by adjusting J_{ij} and D_{ij}. They can be designed by alloying different elements.

The first observation of skyrmions in experiments was reported by Mühlbauer and co-workers. They used small angle neutron scattering (SANS) to measure bulk B20 MnSi [5]. Small angle resonant soft x-ray scattering (R-SoXS) also succeeded in detecting skyrmions in B20 Cu_2OSeO_3 and CoZnMn materials [8, 60]. However, the skyrmions observed by these two methods are in reciprocal space. The skyrmions in real space can be detected by electron microscopy or scanning probe microscopy. The resolution of magnetic force microscopy (MFM) is around 20 nm and thus MFM was used to observe skyrmions in bulk $Fe_{1-x}Co_xSi$ (skyrmion size is around 60 nm) [61] or multi-layer system (skyrmion size is around few hundred nm) [62]. Spin-polarized scanning transmission electron microscopy (SP-STM) was used to detect the atomic size skyrmion in Fe/Ir (111) multi-layers [22]. Lorentz transmission electron microscopy (LTEM) with nanoscale spatial resolution and in-situ magnetic field is the best tool for observing skyrmions directly [42, 44]. There are also some groups who identify skyrmions by magnetic and transport properties [7, 9, 51, 63-66]. From the magnetic hysteresis loop of skyrmion materials, one can observe multi-loops, where the calculated static susceptibility shows several peaks with increasing the magnetic field, indicating magnetic phase transitions. The calculated dMR/dH from magnetoresistance (MR) and Hall resistance can also be used to prove the existence of skyrmions. These magnetic and transport measurements offer an easy and effective way in regular labs to detect skyrmions and will be used in this book.

Due to the fact that skyrmions have good stability, novel dynamic characteristics and can be regulated by magnetic fields, electric fields, currents, etc., they are expected to become a high-density, low-energy, non-volatile information storage and logic operation carrier. To drive skyrmions, one only needs smaller current, which can save energy and decrease the Joule heat. Since magnetic skyrmions were first observed experimentally in 2009, there have been many devices and prototypes based on magnetic skyrmion [5].

1.2 B20 MnSi with magnetic Skyrmions

Magnetic skyrmions were first observed in bulk B20 MnSi materials. B20 MnSi is a member of a class of materials known as cubic helimagnets with a lattice parameter of $a_{MnSi} = 0.4561$ nm at room temperature [63]. MnSi possesses a typical B20 cubic crystal structure (space group P213), which is a distorted rock salt structure with basis vectors (u, u, u) for Mn1 or Si1, ($\frac{1}{2}+u, \frac{1}{2}-u, -u$) for Mn2 or Si2, ($-u, \frac{1}{2}+u, \frac{1}{2}-u$) for Mn3 or Si3, and ($\frac{1}{2}-u, -u, \frac{1}{2}+u$) for Mn4 or Si4. The atomic positions for MnSi are given by $u_{Mn} = 0.137$ and $u_{Si} = 0.845$. A typical X-ray diffraction pattern for bulk polycrystalline MnSi is shown in Figure 1.5. MnSi (210) and (211) result in the strongest reflections. After Rietveld refinement of the XRD data, the lattice parameter a=b=c=0.4559 nm and $\alpha = \gamma = 90°$ can be obtained. Moreover, the secondary $MnSi_{1.7}$ phase grows as a parasite in MnSi compound.

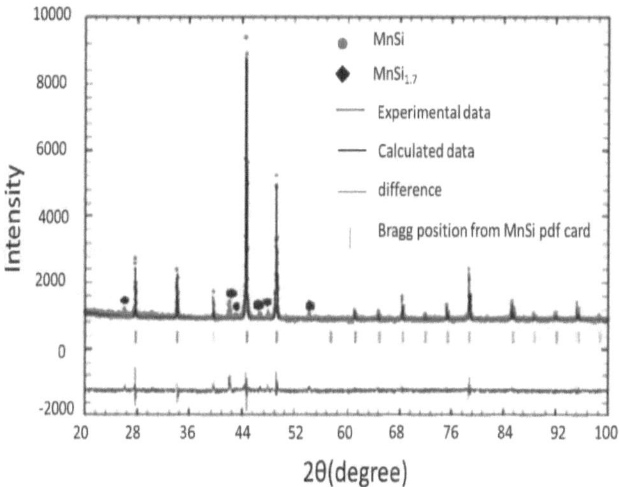

Figure 1.5. The XRD pattern of bulk polycrystalline MnSi. The MnSi$_{1.7}$ phase coexists as a second phase. Reproduced from [63].

As a single crystal, B20 MnSi has no inversion symmetry and hence can have either a left or right handed crystal chirality [67-70]. As we know, chirality plays a critical role in a wide range of systems, from biology and chemistry to condensed matter physics and high energy physics [71]. Bulk MnSi has been studied for decades and become the focus of even more interest after the discovery of the skyrmion lattice in 2009 [4, 5, 19]. This has led to intense investigations in terms of, for example, the Topological Hall Effect (THE) [72], the direct observation of the skyrmion lattice by LTEM [72, 73], collective excitations of different magnetic structures, and effects under uniaxial pressure.

The fabrication of single crystal MnSi compound is very important for the above investigations at the beginning. Crystalline phases are formed by a phase transition from a liquid, gaseous or amorphous state. The main goal in single crystal growth is achieving a reproducible and undisturbed crystallization by studying and controlling the parameters of phase transitions. Thus it is very important to have a detailed knowledge about the equilibrium phase diagrams [74], the growth processes at the phase boundaries and the involved (mass) transport phenomena. The phase diagram is a very important and useful tool to know the phase formation dependent on temperature and composition. The following section provides basic information on the processes during crystallization and crystal growth for MnSi [75].

1.2.1 Crystallization process

It is known that the phase diagram can guide the crystallization process and formation. It is a useful and important tool to describe alloy constitutional phase stability and materials processing. In general, phase diagrams show which phases within a given system are in equilibrium or not, depending on the variables of pressure p, temperature T and molar fraction of each element. An accurate knowledge of the phase diagram allows us to choose the growth method and parameters for stable crystal growth of the wanted compound. Here it is shown how binary phase diagrams can be derived in principle from thermodynamic considerations and the basic types of phase diagrams resulting from these considerations are described [76].

From the phase diagram shown in Figure 1.6 [77], it is known there are three transitions from liquid to solid: congruent reaction, eutectic reaction and peritectic reaction. The reaction from liquid to solid γ phase is a congruent reaction, where the liquid can fully transform into a solid phase with the same composition. The reaction from liquid to two solid $\alpha+\gamma$ or $\gamma+\beta$ phase is the eutectic reaction, where the liquid can fully transform into two solid phases with different compositions. The peritectic reaction refers to a reaction from a liquid and a solid phase to another solid phase. The single phase γ in the phase diagram looks like a line, also called a line compound. For most B20 compounds, they are line compounds. However, some B20 compounds can not be found from the phase diagram, indicating that these phases can not be formed in an equilibrium conditions. B20 MnSi is a line compound and can be formed at the composition ratio of Mn and Si close 1:1. From the phase diagram, it is known that B20 MnSi can be formed at ambient pressure.

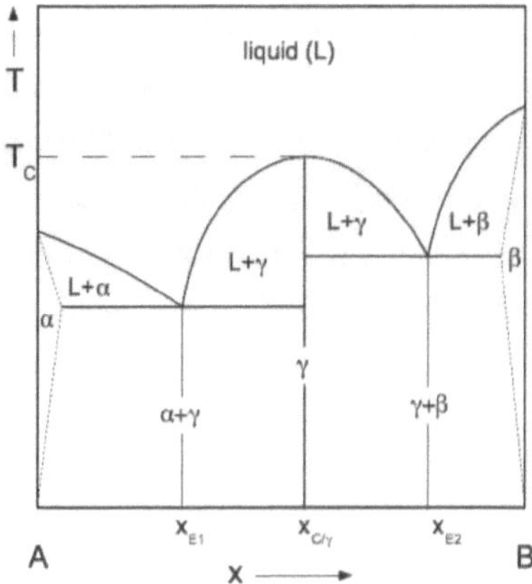

Figure 1.6. Basic types of phase diagrams for binary systems with congruent reaction and eutectic reaction. Reproduced from [77].

Nucleation and growth are the most important two stages for crystallization process [78]. In the initial nucleation stage, a small nucleus is formed and contains the newly forming crystal. Nucleation is relatively slow because the initial crystal components must meet with the correct orientation and placement to adhere to each other and form the crystal. After the successful formation of a stable nucleus, the crystal growth starts in which free particles (atoms or molecules) adsorb to the nucleus and spread the crystalline structure outward from the nucleation site. This process is much faster than nucleation. The reason for such rapid growth is that true crystals contain dislocations and other defects that act as catalysts for particles to attach to the existing crystalline structure. In contrast, perfect crystals (without defects) would grow extremely slowly. In all these two processes, the driving force is the source to promote the reaction progress.

Driving force

The driving force for crystallization per volume ΔG_v is defined as the difference between the free energy of the liquid and crystalline state [79]. It plays the major role for crystallization process. It is obvious that the lateral growth mechanism is found when any region in the surface can reach a metastable equilibrium in the presence of a driving force. It will then tend to remain in such an equilibrium configuration until a step is passed. Thereafter, the configuration will be identical except that each part of the step will have advanced by the step height. If the surface cannot reach equilibrium in the presence of a driving force, it will continue to advance without waiting for steps to move laterally. Thus, Cahn concluded that the distinguishing feature is the ability of the surface to reach an equilibrium state in the presence of the driving force [80]. He also concluded that for every surface or interface in a crystalline medium, there exists a critical driving force, which, if being exceeded, will enable the surface or interface to advance normal to itself, and, if not being exceeded, will require the lateral growth mechanism. Therefore, for sufficiently large driving forces, the interface can move uniformly without a heterogeneous nucleation or screw dislocation mechanism. What constitutes a sufficiently large driving force depends on the diffusivity of the interface, so for extremely diffuse interfaces this critical driving force is so small that any measurable driving force exceeds it. For sharp interfaces, on the other hand, the critical driving force is very large and most growth will occur through the lateral step mechanism.

In a typical solidification or crystallization process, the thermodynamic driving force is dictated by the degree of supercooling. The free energy, which is a criterion for a solution reaction, the standard thermodynamic equation for free energy calculations is shown as follows [80]:

$$\Delta G_V = \Delta G^O + RT \ln Q \tag{6}$$

where ΔG^0 is the standard free energy, R is the gas constant, T is the temperature and Q is the activity, respectively. Nucleation is only thermodynamically favoured when $\Delta G v < 0$.

1.2.2 Phase diagram of Mn-Si binary compound

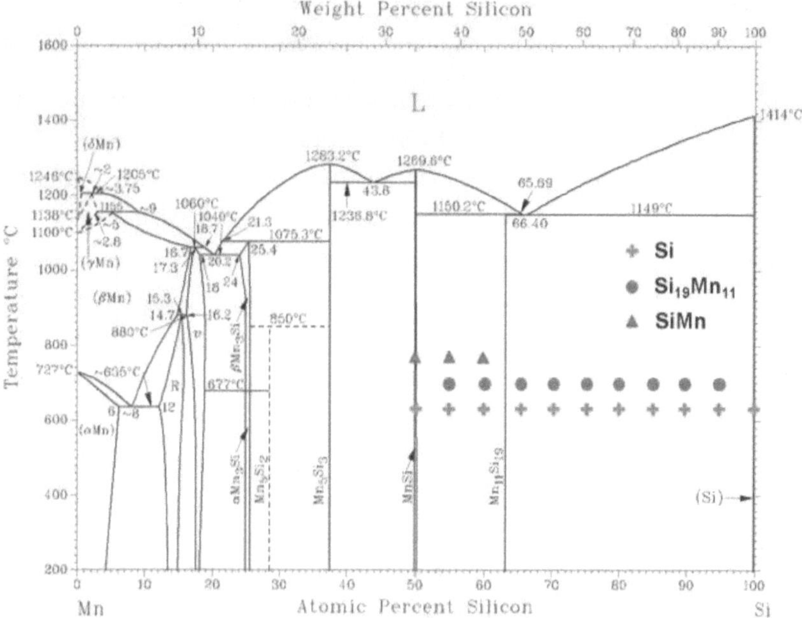

Figure 1.7. The equilibrium phase diagram of Mn-Si system. The red triangle, blue dot and green cross represent the B20-MnSi, MnSi₁.₇ and Si phase. Single MnSi phase can only exist at 50 at.% Si, however MnSi₁.₇ appears in a much broader window. Reproduced from [81].

The equilibrium phase diagram of the binary Mn-Si alloy is shown in Figure 1.7 [81]. As the Si composition is lower than 18.75 at.%, there is no Mn-Si compound and the Si atoms are at the substitution site or in the interstitial site. When the Si concentration is increased to 20.2 at.%, there is a eutectic reaction from the liquid to two solid phases: v and β-Mn₃Si at 1040 °C. When the Si concentration varies from 25 to 25.6 at.%, single Mn₃Si phases are stable. The high temperature β-Mn₃Si phase converts to the room temperature α-Mn₃Si phase at about 677 °C, which is known as a polymorphic reaction. When the Si concentration is increased to 37.5 at.%, the line compound Mn₅Si₃ can be formed from the liquid by congruent reaction at 1300 °C. Mn₃Si can also be formed by a peritectic reaction from liquid phase + solid Mn₅Si₃ phase to the Mn₃Si solid phase. The eutectic reaction of liquid to solid Mn₅Si₃ and B20-MnSi can occur at 45.6 at.% Si at about 1234 °C. The single B20-MnSi phase can form only at 50 at.% Si by the congruent reaction at 1276 °C. With further increase of Si concentration, there are other compounds called higher manganese silicides, which have many chemical formulas, such

13

as: Mn_4Si_7, $Mn_{11}Si_{19}$, $Mn_{15}Si_{26}$ and $Mn_{27}Si_{47}$ and are commonly written as $MnSi_{1.7}$. A single $MnSi_{1.7}$ phase can be formed from 63 to 63.64 at.% Si. As shown in the phase diagram, the region between 50 and 63at.% Si is the mixture of B20-MnSi and $MnSi_{1.7}$.

B20 MnSi with non-centrosymmetric cubic structure accommodates skyrmions and has a transition temperature at around 29.5 K [5]. Higher manganese silicides are reported as a weak itinerant magnet with a saturation magnetization of 0.012 $\mu B/Mn$ [82], which is 20 times lower than B20 MnSi. However, as a semiconductor with narrow band gap of 0.4-0.7 eV, higher manganese silicides can be an excellent candidate for thermal electric materials and near-infrared detectors [83-86]. Mn_3Si crystallizes cubically with a DO3 structure that exhibits non-collinear antiferromagnetic behavior at a Neél temperature (T_N) of 25 K [87-89]. Orthorhombic Mn_5Si_3 also exhibits non-collinear antiferromagnetic behavior below a T_N of 68 K and collinear antiferromagnetic behavior below a T_N of 99 K [90, 91].

1.2.3 Bulk B20-MnSi

Melt growth is the commercially most important method of crystal growth. Growth from melt can be further divided into several different techniques [76, 92]: a) Bridgmann method; b) Czochralski method; c) Vernuil method; d) zone melting method; e) Kyropoulos technique; and f) skull melting. The Czochralski method and the floating zone melting method are the most common methods for fabricating bulk MnSi single crystals. The Czochralski method is named after the Polish scientist Jan Czochralski, who invented the method in 1915 while studying the crystallization rates of metals [93]. He made the discovery by an accident: instead of dipping his pen in his inkwell, he dipped it in molten tin and drew a tin thread, which later turned out to be a single crystal. The vertical floating zone method was first introduced by Kec and Golay in 1953 for the crystallization of high-purity silicon [94]. In this growth technique, as the name implies, a narrow melt zone is created by local heating, optically or inductively, between two vertically mounted, free-standing rods.

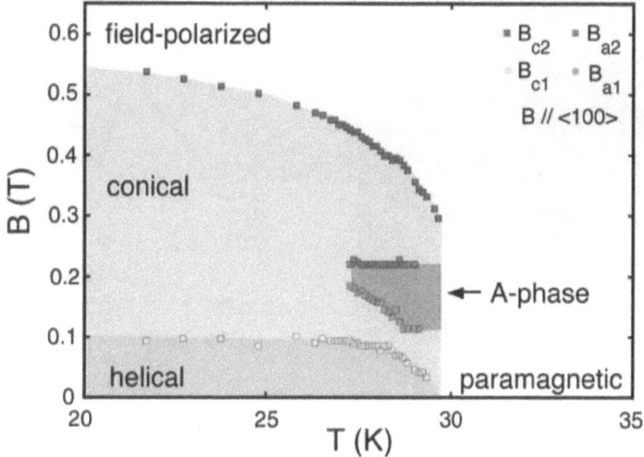

Figure 1.8. Magnetic phase diagram of bulk MnSi crystals near the Curie temperature T_C=29.5 K. The A-phase is the magnetic skyrmion phase. Reproduced from [5].

Due to its special crystalline structure, B20 MnSi exhibits interesting and complicated magnetic properties. All magnetic phases of bulk B20 MnSi are summarized in a magnetic field-temperature (B-T) phase diagram measured by reciprocal space SANS in Figure 1.8. When bulk MnSi cools from room temperature with zero magnetic field, the paramagnetic to helimagnetic transition arises at T_C = 29.5 K. Below the Curie temperature, the spins rotate in the plane perpendicular to the propagation direction of the helix. The helices are left-handed, with a length of around 18 nm. Moreover, applying a magnetic field at around B_{c1} = 0.1 T induces a transition from helix to conical phase, where the wave vector k_H of the helices aligns along the field direction and the spins start to tilt towards B until the ferromagnetic state is reached at high fields above 0.55 T. The critical field of this ferromagnetic transition, i.e. B_{c2}, drops with increasing temperature due to the decrease of anisotropy. In the regime of the conical phase, the phase diagram includes the first example of a new form of magnetic order composed of topologically distinct spin solitons at around Curie temperature, namely skyrmion, which is illustrated in Figure 1.9 with the mark of 'A phase'. Skyrmions can exist from the critical field B_{a1} to B_{a2} and from 26.5 K to Curie temperature.

Later on, researchers have found that the Curie temperature of bulk MnSi can be increased by applying stress [95] or by thinning the crystals [96]. The temperature and field range in which magnetic skyrmions are stable can also be enlarged.

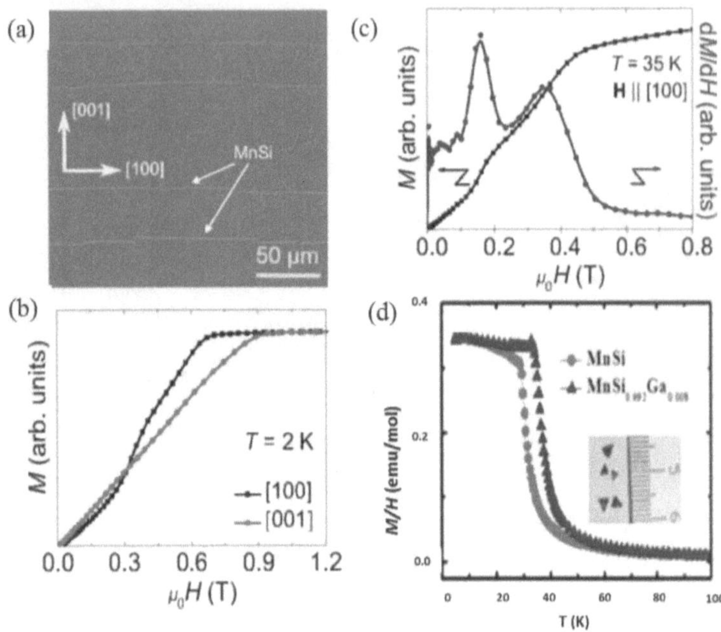

Figure 1.9. (a) The micrograph of MnSi lamellae embedded into the nonmagnetic Mn$_{11}$Si$_{19}$ (MnSi$_{1.7}$) matrix. (b) magnetization as a function of the field at T = 5 K; (c) Metamagnetic transitions at 35 K; (d) Temperature (T) dependent , M divided by the field, H, for MnSi and MnSi$_{0.992}$Ga$_{0.008}$ measured at H=1 kOe. Inset: pyramidal shaped crystals grown using Ga flux. Reproduced from [97] and [98].

Sukhanov et al. prepared bulk MnSi$_{1.7}$ single crystal and found B20 MnSi with a line shape embedded into MnSi$_{1.7}$ matrix shown in Figure 1.9 (a). From magnetic hysteresis loops (MH) curves measured at 2 K along [100] and [001] directions in Figure 1.9 (b), the easy axis is in the [100] direction and multi-hysteresis is found in this direction, indicating the existence of skyrmions [97]. As shown in Figure 1.9 (c), they calculated dM/dH from the MH curve measured at 35 K to investigate the transition fields for skyrmions. They also proved the skyrmion formation by SANS. Dhital et al. prepared MnSi and MnSi$_{0.992}$Ga$_{0.008}$ single crystal by the Bridgeman method. They found that the Curie temperature is 29.5 K and the substitution of Si by Ga can improve the Curie temperature to 33.5 K shown in in Figure 1.9 (d) [98].

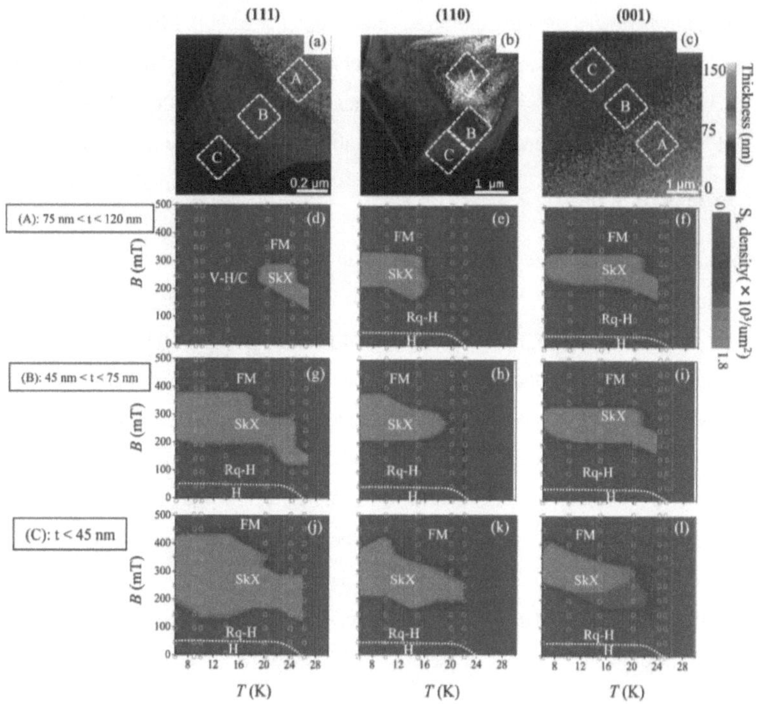

Figure 1.10. Magnetic phase diagrams obtained for wedge-shaped (111), (110), and (001) MnSi thin plates. (a)–(c) Thickness maps of the (a) (111), (b) (110), and (c) (001) plates, respectively. (d)-(f): phase diagrams for (A) thick regions; (g)-(i) for (B) intermediate-thickness regions between 45 nm and 75 nm; (j)-(l): for (C) thin regions with thickness below 45 nm. SkX stands the magnetic skyrmion phase. Reproduced from [96].

After the discovering of magnetic skyrmions in bulk MnSi by reciprocal space neutron scattering, Yu et al. also studied the formation of skyrmions in thin plate samples obtained from bulk crystals by real space LTEM [96]. By this approach, they could investigate the influence of plate thickness on the magnetic skyrmions. The results are shown in Figure 1.10. According from the phase diagram dependent on plate thicknesses, it is known that the shape anisotropy based on the long ranged magnetic dipolar interaction or finite size effects may play an important role on the skyrmion formation window. Additionally, the surface anisotropy dominates with increasing the ratio of surface to volume atoms by decreasing the thicknesses. The temperature range for stabilizing skyrmion is enlarged. However, these systems do not experience uniaxial strain from substrates, as in the case of epitaxially grown thin films. Thus, some groups investigated the pressure on skyrmion stability of bulk MnSi single crystals and the skyrmion range can be enlarged under external pressure. Therefore, for thin films a different magnetic behavior than bulk samples or thin plates is expected, and these differences can be attributed solely to the additional anisotropies: strain or/and interface anisotropy.

1.2.4 B20-MnSi thin film

Butenko et al. made some theoretical predictions on the stability of skyrmion lattices in a strained epitaxial thin film [99], where the strain induces the change of magnetocrystalline anisotropy in the film. They predicted that uniaxial anisotropy suppresses the helical and conical phases when an in-plane magnetic field is applied, leading to a much broader skyrmion stability region than seen in bulk crystals. The potentially broader skyrmion window, combined with the modern integration silicon-based electronics, makes the growth of thin films of MnSi helimagnets on silicon substrates as an attractive research area. Up to now, the most prominent skyrmion phase in bulk B20 materials can only be realized below room temperature and at finite magnetic fields. For practical applications, it is necessary to have a stable skyrmion phase at room temperature and wide magnetic field range. Thin films are a good avenue for achieving these goals, since the size effect can increase the range of temperatures and magnetic fields within which skyrmions are stable.

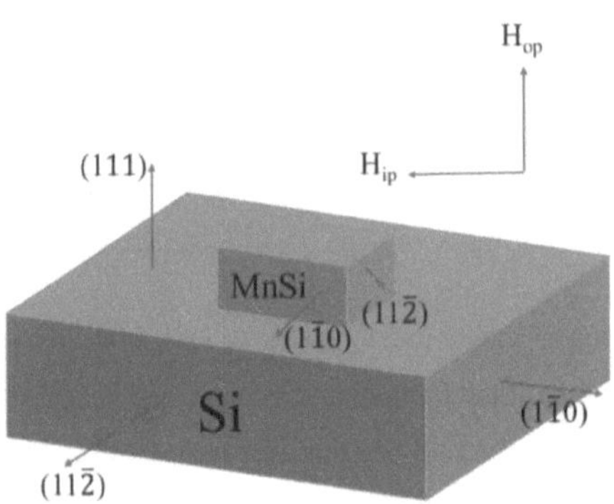

Figure 1.11. A schematic of the epitaxial relationship MnSi[1 $\bar{1}$ 0]||Si[11$\bar{2}$] for MnSi(111) on a Si(111) substrate. The lattice mismatch for this geometry is -3.1%. Reproduced from [100].

Epitaxial thin films of B20 materials have drawn considerable attention because of their improved Curie temperature and skyrmion region [101]. (111)-MnSi thin films on Si(111) substrates are the most studied B20 thin films [100, 102-106]. MnSi(111) thin films are usually grown epitaxially by thermal deposition using either solid phase epitaxy (SPE), where Mn or a Mn/Si multilayer is annealed to form MnSi films or by molecular beam epitaxy (MBE) [72, 102-104, 106-108], where Mn and Si are co-deposited simultaneously followed or together with annealing at 400 °C. All these processes are in ultra-high vacuum conditions. In the MBE case, few monolayers of Mn have to be annealed on the Si substrate to form a MnSi seed layer before co-deposition of Mn and Si. MnSi typically grows along its <111> direction on Si(111) substrates. To overcome the large lattice mismatch between MnSi with a_{MnSi} = 4.561 Å and Si with a_{Si} = 5.431 Å, MnSi grows with a lattice rotation of 30^0 and its $<1\bar{1}0>$ is parallel to Si $<11\bar{2}>$ of the Si substrate. This orientation relationship of the MnSi film and Si substrates is shown in Figure 1.11. This orientation relation leads to tensile strain of $(a_{MnSi}-a_{Si}\cos(30^0))/a_{Si}$ = −3.1%. With respect to the lattice mismatch, the rotation of 30^0 can be left or right. Thus the reported MnSi thin films have twins between 30^0 rotations of left and right, indicating the double chirality.

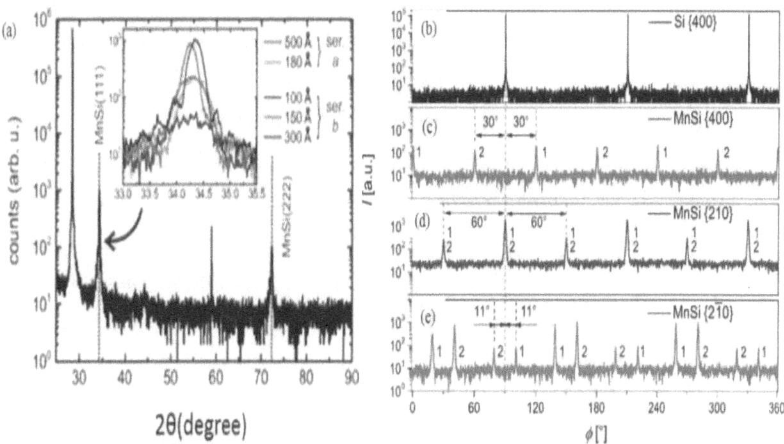

Figure 1.12. X-ray diffraction patterns of MnSi films on Si(111). An overview 2θ−ω scan for the 300-Å-thick sample shows the peaks for the MnSi film and Si substrate. The inset shows the enlarged MnSi(111) peaks for the films prepared under different conditions. Azimuthal ∅-scans of asymmetric Si and MnSi XRD reflections. Two different MnSi domains are labeled by 1 and 2. (b) Si {400}, (b) MnSi {400}, (d) MnSi {210}, and (e) MnSi {2$\bar{1}$0}. Reproduced from [108] and [109].

The most studied B20 MnSi thin films are grown on Si(111) substrates. As shown in Figure 1.12 (a), Figueroa et al. measured XRD for epitaxial MnSi films [108]. The MnSi(111) and (222) Bragg peaks can be detected, indicating epitaxial growth. With increasing the thickness of MnSi films, the Bragg peaks become sharp, indicating an excellent crystal quality. Trabel et al have investigated the orientation relation of MnSi film and Si substrate, as shown in Figure 1.12 (b-e) and found that the MnSi rotates 30^0 with respect to the substrate to reduce the lattice mismatch [109]. They also found 6 peaks for MnSi(400) while only 3 peaks for Si(400), indicating that MnSi film has crystalline twins. In the {2 10} direction, since the left hand and right hand rotated twins stay at the same azimuthal position, thus no extra peak is detected. However, in the {2 $\overline{1}$ 0} direction, domains 1 and 2 locate at different $\emptyset$ degrees and the result shows 12 peaks (see Figure 1.12(e)).

Daisuke et al. succeeded in visualizing the domain structure of the chirality and axis orientation of a MnSi thin film by using a combination of transmission electron microscopy and x-ray reflectivity measurement [69]. Two types of domains were found for MnSi films grown on a Si(111) substrate. Since the magnetic skyrmion cannot move across the chiral domain boundary, the characterization and understanding of chiral twin structure in the film are important for exploring skyrmion behavior. The determination of crystal chirality is very important to the determination of spin chirality, which is one-to-one correspondence to the crystal chirality. For the bulk B20 MnSi, it is known that the left-hand chirality exists due to the absence of substrate and it is easy to determinate the chirality in bulk materials [5]. However, B20 MnSi thin films always contain double chirality due to the twin structure. In the case of the chiral twin state, a method for the spin helicity determination using a probe with a spatial resolution worse than the typical size of domains such as neutron scattering cannot determine the chirality for each single domain. Thus to analyse the chirality and a chiral domain in thin films, a method with nanometer resolution like transmission electron microscopy (TEM) and/or electron backscattering diffraction (EBSD) is required. In theory, EBSD can also identify the crystal-chirality domain by a comparison with simulation using dynamical scattering theory. Kikuchi patterns from EBSD measurement are also possible to conclude the crystal chirality twins.

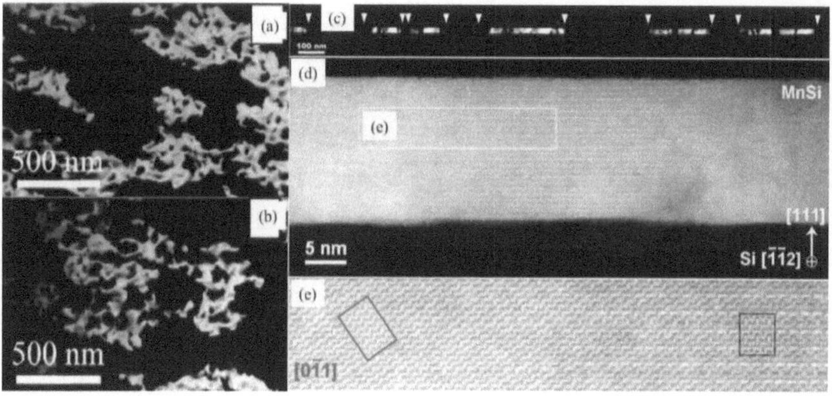

Figure 1.13. (a) and (b) show plane-view TEM dark-field images of a 17.6-nm-thick MnSi film. The complementary (01̄2) and (1̄02) reflections were used in (a) and (b), respectively, which have opposite contrast for opposite crystal chiralities. (c) Dark field TEM overview. (d) HAADF-STEM image. (e) Enlargement from (d) showing one of the twin domains. Simulations of both regions are shown inside the red and red/blue rectangles. Reproduced from [105] and [109].

Karhu et al. first reported the chiral domain structure in a MnSi thin film by the contrast in dark-field TEM images shown in Figure 1.13 (a) and (b) [105]. However, this report included an error for indexing and a misunderstanding. To determine the chiral structure, a quantitative comparison is needed in the case of TEM observation. However, no evidence was provided in the report that the observed dark-field TEM images should be attributed to the enantiomorphic twins before. In principle the positions of diffraction spots are not related to the crystal chirality. Convergent-beam electron diffraction (CBED) is another tool for investigating the chirality. Nonetheless, it is difficult to distinguish the handedness of a MnSi thin film grown on a substrate with [111] orientation because the intensity of the electron diffraction from the thin film is strongly affected by the substrate. Thus the thickness of film must be thicker than ten nanometers to detect a significant difference between the two enantiomers. In a later study, Trabel et al. reported the chiral domain structure with 200 nm combining TEM (Figure 1.13 (c)) and XRD (Figure 1.12 (b-e)) measurements on a 17 nm B20 MnSi film [109]. One of the twin domains and a region in which both overlap are marked red and red/blue rectangles, respectively, in Figure 1.13 (e). They found the chiral domain structure with rotated crystal axes by 30 degree respect to the crystal axes of Si.

The magnetic properties of MnSi films have also been well investigated. Figure 1.14 (a-c) shows the magnetization of MnSi films with different thickness [106]. All these three films with different thicknesses show a similar saturation magnetization of around 163 kA/m. With increasing the thickness from 11.6 nm (Figure 1.14 (a)) to 26.7 nm (Figure 1.14 (c)), two transition fields are emergent since the skyrmion size in MnSi is around 18 nm. To better observe the transitions of different magnetic structure, Wilson et al. did the calculation of dM/dH from MH curves shown in Figure 1.14 (d). The peak of dM/dH curves indicates these critical fields of different magnetic phases. They also measured the out-of-plane strain by XRD and in-plane strain by TEM technique as a function of thickness. They found the in-plane and out-of-plane strain in these samples prepared by different methods have a big difference. The MBE grown samples show a slightly higher in-plane strain between 0.5% and 1.2% and slightly higher absolute values of out-of-plane strain between 0.25% and -0.5%. They concluded that the change of strain affects the Curie temperature (Figure 1.14 (e)) which is found to be around 43 K, much higher than the value for bulk MnSi.

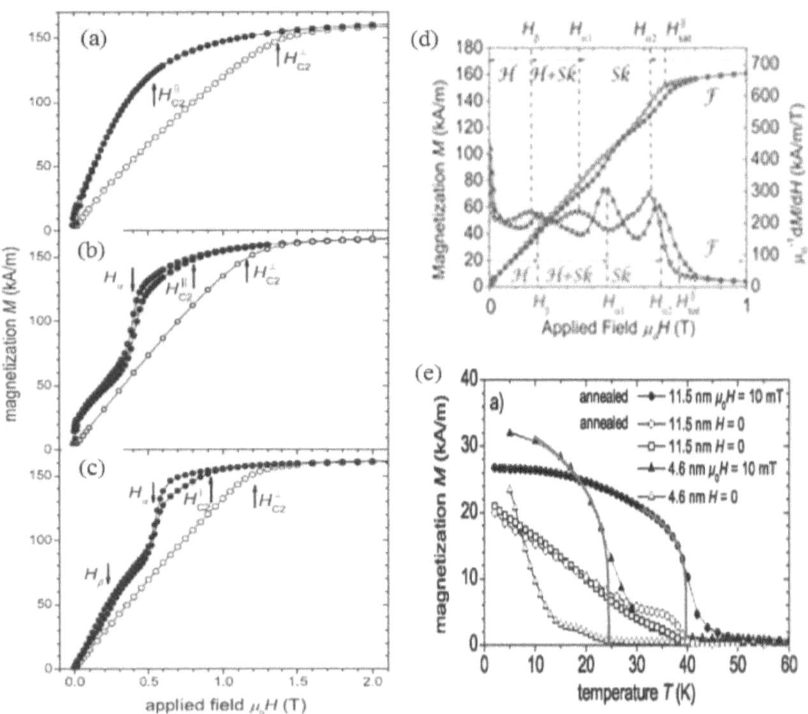

Figure 1.14. M-H curves of MnSi thin films with an in-plane field applied along [110] (filled points) and out-of-plane field along [111] (open points) measured at 5 K. The film thicknesses are d = 11.6 nm (a), 17.6 nm (b), and 26.7 nm (c). M-H curves (squares) and dM/dH (triangles) for d = 26.7 nm film for increasing (red filled points) and decreasing (blue open points) magnetic field at 15 K (d). The elliptic skyrmion (Sk), helicoid (H), elliptic cone (C), and ferromagnetic (F) regions are labeled. (e) The open symbols show the remanent magnetization of MnSi films, measured on warming the sample. The filled symbols show the field-cooled magnetization together with power-law fits, shown by the thick solid lines. The data labeled with 'annealed' corresponds to a sample that was heated ex situ at 400 °C for 1 h. Reproduced from [102, 106].

Meynell et al. investigated the magnetic skyrmion phase in MnSi films. They plotted the phase diagram for a MnSi film of around 26 nm by using magnetometry measurement, planar Hall effect (PHE) and polarized neutron reflectometry (PNR) as shown in Figure 1.15 [110]. With increasing the magnetic field to H_β, the magnetic state starts to change from Helicoid to Helicoid+Skyrmion. With further increasing the field to critical filed $H_{\alpha 1}$, all helicoids change into whirling skyrmion, which can be stabilized at the range from $H_{\alpha 1}$ to $H_{\alpha 2}$. Compared with bulk MnSi, the temperature and magnetic field window for stabilizing skyrmion are enlarged. The Curie temperature of MnSi thin films are increased to around 43 K. Skyrmions can be stabilized not only in the range near the Curie temperature but also in the much lower temperature shown in Figure 1.15. Moreover, due to the anisotropy induced by strain, B20 MnSi thin film can only host skyrmion in plane.

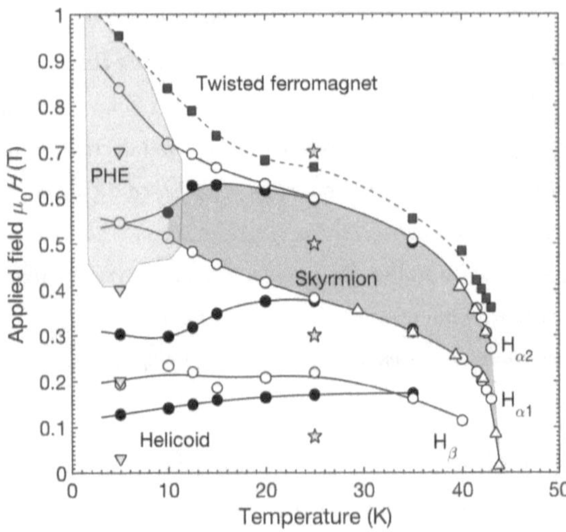

Figure 1.15 The phase diagram for a MnSi/Si(111) film with H\[1$\bar{1}$0]. The filled (unfilled) circles correspond to peaks in dM/dH measured in decreasing (increasing) magnetic fields for sample of 26.7 nm thick. The blue squares are minima in d^2M/dH^2 for increasing magnetic fields [101]. The polarized neutron reflectometry (PNR) experiments were performed in successively decreasing magnetic fields at the (H, T) points shown by the yellow stars and by the yellow triangles (Ref. [111]). The cyan-colored area taken from Ref. [112] shows the region where a drop in the planar Hall effect (PHE) signal is observed in a similar d = 26-nm sample measured in a decreasing field. Reproduced from [110].

Regarding the mixture with other Mn-silicide phases, Karhu et al. found that the formation of $MnSi_{1.7}$ precipitates with a diameter of up to a few hundred nm is difficult to avoid. One typical $MnSi_{1.7}$ precipitate is shown in Figure 1.16 [105]. As shown in Figure 1.7 of the phase diagram, it is found that the $MnSi_{1.7}$ precipitate can coexist with B20 MnSi due to its lower crystalline temperature and the endless supply of Si from substrates. In the work on higher manganese silicides, they also found B20 MnSi always precipitates as a second phase and deteriorates the thermal properties of higher manganese silicides. Karhu et al. found that B20 MnSi films with volume fractions of $MnSi_{1.7}$ up to 11% still show a similar T_c of 41.4–44.0K. They concluded the precipitates of $MnSi_{1.7}$ do not affect the magnetic properties of MnSi films. However, the preparation of thin film materials with high purity is a prerequisite for preferred day's information technology. Therefore, how to separate $MnSi_{1.7}$ impurities from MnSi film and how does this MnSi$_{1.7}$ impurities affect the magnetic properties and the skyrmion phase are very important prerequisites for condensed matter physics and future applications.

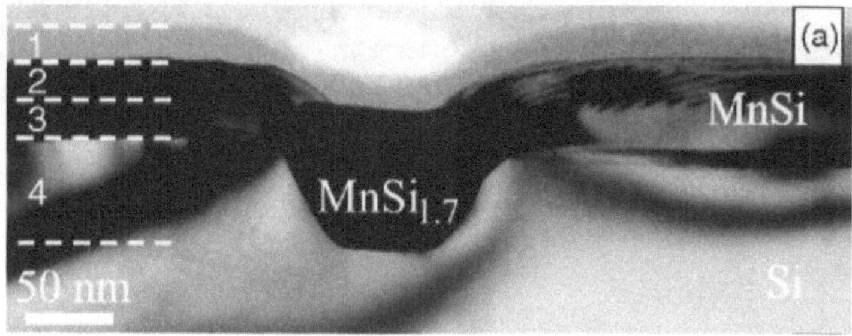

Figure 1.16. Cross-section bright-field TEM image of the 39.5-nm-thick MnSi film showing one of the MnSi$_{1.7}$ precipitates. Reproduced from [105].

As we discussed, B20 MnSi films were mostly prepared on Si(111) substrates. Due to the symmetry, the growth of MnSi films on Si(100) substrates is difficult and has not been reported so far as our knowledge.

1.2.5 B20-MnSi nanowire

It is known that the skyrmion lattice in bulk MnSi appears only in a small region (known as the A phase) of the magnetic field–temperature (H–T) phase diagram in bulk samples, while in thin films, the skyrmion phase is much more robust. It is of great interest to investigate the properties of quasi-1D MnSi nanowires (NW) to see the effect of an extra dimensionality limitation [6, 73, 113-118]. The persistence of the skyrmion ordering in one-dimensional MnSi nanowires may be very promising for spintronics applications as the magnetic domains and individual skyrmion could be manipulated with small currents. Recently, several reports have appeared about the successful synthesis of high-quality MnSi nanowires via chemical vapor deposition, using MnCl$_2$ or Mn vapor as a precursor gas.

Du et al. prepared MnSi NW and investigated the formation of skyrmions by transport measurement [115, 117]. They also found an expanded skyrmion window and improved Curie temperature up to around 32 K. A single wire with a smooth (111) surface and a diameter of 40 nm (larger than the theoretical size of skyrmion in MnSi materials) was measured by scanning electron microscopy and shown in Figure 1.17 (a). They used the four-probe method to measure the resistance and magnetoresistance (MR) of individual NWs at 0 and 5 kOe, respectively. There is an obvious valley (Figure 1.17 (b)) at around 29.5 K, indicating the Curie

25

temperature, which is consistent with bulk MnSi and lower than MnSi film. They also measured MR of thicker NW with a diameter of 410 nm. Interestingly, this thicker NW has a higher Curie temperature up to 32 K, when the anomalous behavior in magnetoresistance disappears shown in Figure 1.17 (c). By calculating the derivative dMR/dH (Figure 1.17 (d)), the transition fields of the different magnetic phases are obtained. It is clear that the skyrmion phase is further stabilized in MnSi NWs compared to bulk MnSi.

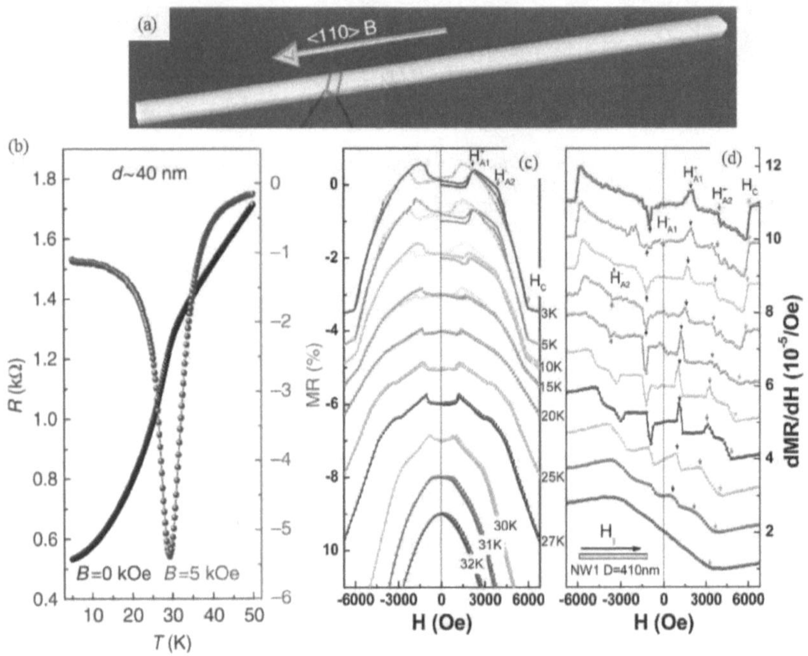

Figure 1.17. (a) A typical scanning electron microscopy (SEM) image of a MnSi NW grown along <110>. (b) Temperature dependence of the resistance (black) and magnetoresistance (MR; coloured dots) at 0 and 5 kOe, showing a TC of 29 K. (c) Typical variation of the normalized MR as a function of external magnetic field H∥ at various temperatures. (d) The derivative dMR/dH for the same data in (c). The peak at HA1 indicates the first-order phase transition from helimagnetic to skyrmion states. Hc is defined from the kink as the transition from the conical phase to FM. Each curve is shifted vertically for clarity. Reproduced from [115, 117].

In these nanowires, higher manganese silicides are also found to coexist with MnSi. By comparing MnSi bulk crystals, thin films and nanowires, it seems that MnSi thin films have the highest Curie temperature of around 43 K and the broadest skyrmion region. To achieve the application of skyrmion in modern electronics and spintronics, thin films are also the most appropriate avenue. However, we must solve the problems of growing MnSi thin films: a) how to avoid or minimize the parastic $MnSi_{1.7}$ phase, b) if one can prepare MnSi films on Si(100) substrates and c) how to obtain homochiralical MnSi films.

1.3 Fast annealing method

As discussed in previous sections, one critical problem in the growth of MnSi films is the presence of the parasitic $MnSi_{1.7}$ phase [105]. I hypothesize that this problem can probably be solved by applying a rapid reaction a higher temperature in contrast to the slow solid-phase reaction at low temperature. This idea is borrowed from the preparation of Zr-based bulk metallic glass alloys [119]. In this section, I will give a brief review about the development and application of rapid heating or flash annealing on various material classes.

For heating treatment, the heating rate, final annealing temperature, annealing time and cooling rate are important parameters. In flash annealing process, we care more about heating/cooling rate and annealing temperature. Different from traditional annealing, flash annealing is a non-equilibrium process, which can form some non-equilibrium phases and selected microstructures. Flash annealing method can refine the grain size, which can improve mechanical properties, like, strength, hardness, or magnetic properties, like magnetic permeability or saturated magnetization. Fast annealing can also remove the secondary phase and extend phase formation window of the desired phases.

In metallurgy, the thermal processing parameters, particularly the heating and cooling rates play a nontrivial role. Fast annealing is a useful method to select desired phases. Figure 1.18 is the continuous heating transformation diagram of CuZr based bulk metallic glass. There are three possible phases by annealing metallic glass. B2-CuZr is one ductile phase, which enhances the mechanical properties. However, $Cu_{10}Zr_7$ and $CuZr_2$ are brittle, which deteriorate the mechanical properties. However, traditional annealing methods always form $Cu_{10}Zr_7$, $CuZr_2$ and B2-CuZr together due to the similar nucleation energy since $Cu_{10}Zr_7$ and $CuZr_2$ are low-temperature thermal-equilibrium phases. Kosiba et al. have used flash annealing with controlled heating rate and succeeded in separating these phases. From Figure 1.18, we can see $Cu_{10}Zr_7$, $CuZr_2$ and B2-CuZr phases appear together when the heating rates is below the critical heating rates (16 K S^{-1}). Upon increasing the heating rates, the glass transformation temperature

T_g and crystalline temperature T_x increase. By increasing the heating rate, the width of the supercooled liquid region can be greatly broadened.

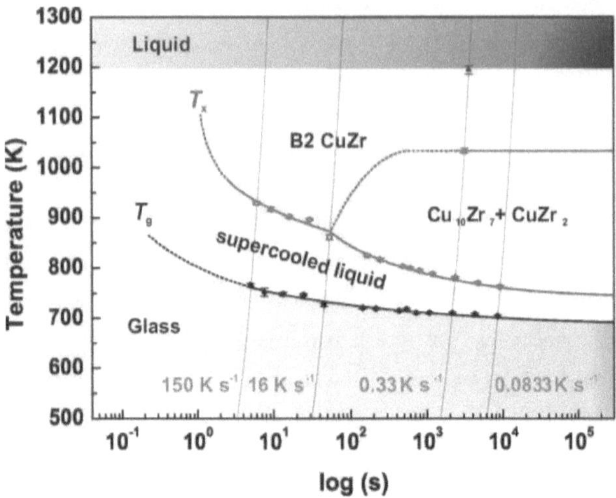

Figure 1.18. Continuous heating transformation diagram for glassy Cu₄₄Zr₄₄Al₈Hf₂Co₂ heated at rates between 0.08 and 150 K s⁻¹. At heating rates larger than 16 K s⁻¹, the crystallization process is changed from eutectic to polymorphic. A B2 phase CuZr alloy can be obtained. Note that at smaller heating rates, this B2-CuZr alloy always transforms to a low temperature thermal equilibrium phases including Cu₁₀Zr₇ and CuZr₂. Figure is from Ref. [120].

Controlling the heating and cooling rate during the materials growth provides a promising option to separate different phases. It can be understood in the following way. The relationship of the nucleation activation energy Q, heating rates Ø and crystallization temperature T_p can be expressed by the Kissinger equation [121, 122],

$$\ln\left(\frac{\emptyset}{T_p^2}\right) = -\frac{Q}{RT_p} + A \qquad (10)$$

where R is the gas constant, and A is a constant, respectively. For a particular crystal phase, when increasing the heating rates Ø, the crystallization temperature will be shifted to higher temperatures and the crystallization temperatures for different phases can be separated, as schematically shown in Figure 1.19. In the situation of ultra-fast heating, temperature increases too fast so that some phases cannot nucleate. As shown in Figure 1.18 for CuZr-based metallic glasses, by increasing the heating rate above 16 KS⁻¹, ductile B2 phase has been selectively formed and brittle phases ($CuZr_2$ and $Cu_{10}Zr_7$) are suppressed. The fast heating and cooling rates make the system-temperature over or below the crystalline temperature for brittle phases in a very short time. There is no nucleation time for the brittle phases due to the transient heating and cooling processes. This approach also works well for amorphous Fe-based metallic glasses.

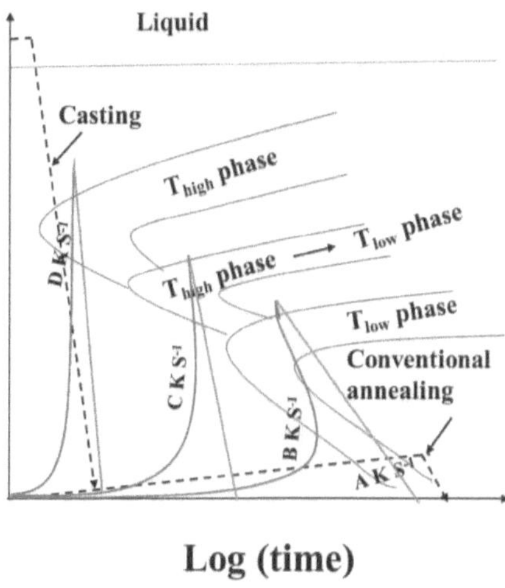

Figure 1.19. Continuous Heating Transformations (CHT) diagram. The red curves stand for annealing processes at different heating rates and the green curves stand for cooling processes. The A, B, C, D and E stand for the increasing heating rates. T$_{low}$ *phase and T*$_{high}$ *phase stand for the phase formed at low temperature and high temperature, respectively. When the temperature is over critical heating rate D, low temperature phase can be suppressed. If cooling process is also very fast, like the left process, the low temperature phase can be avoided.*

This indeed highly motivates that by controlling the heating and cooling rate during the reaction between transition metal and Si, we probably can get single-phase B20-phase monosilicides thin films, which are not possible by conventional thin film preparation as we shown in the previous section. Compared with flashed Zr-based bulk metallic glasses shown in Figure 1.18, MnSi is one high temperature phase like B2 CuZr and $MnSi_{1.7}$ is a low temperature phase like $CuZr_2$ and $Cu10Zr_7$.

1.4 Objectives and the structure of the book

As a canonical example of Skyrmion hosting materials, B20 MnSi has been in focus for two decades. Although bulk B20-MnSi compounds are relatively easy to fabricate by rod-casting furnace or Czochralski method at ambient pressure, the skyrmion can only exist in a very narrow temperature and magnetic field window. It would be desirable to have B20-MnSi thin films that can stabilize skyrmions in a broader window, and thin film materials are more suitable for integrated spintronic device. It is also known that Si(100) has better compatibility with complementary metal-oxide-semiconductor (CMOS) technology [123-126]. However, there are no reports about B20 MnSi on Si(100) substrates as our knowledge. Furthermore, the growth of B20 MnSi on Si(111) substrates always includes the precipitation of second phase $MnSi_{1.7}$.

Accordingly, the present book focuses on the fabrication of MnSi films on both Si(111) and Si(100) substrates by flash lamp annealing. The detailed structural, magnetic and magneto-transport characterization of the obtained MnSi films will be discussed. The book is arranged in the following way:

Chapter 1 is an introduction of the state-of-the-art for the investigated material. And Chapter 2 includes the brief description for involved experimental techniques, particularly for sample preparation.

The third chapter focuses on the growth and characterization of Mn/Si regrown layers on Si(100) substrates. To obtain B20 MnSi, we deposited Mn with thicknesses from 7 nm to 30 nm and annealed these samples with different annealing parameters. Afterwards, the phase compositions of regrown layers are checked by X-ray diffraction (XRD), Raman and TEM. The formation of skyrmion and the spin texture is checked by magnetic and transport measurements.

The fourth chapter focuses on the separation of B20 MnSi and $MnSi_{1.7}$ phases for films grown on Si(111). By tuning the annealing parameter, the annealing temperature can be adjusted. Afterwards, single B20 MnSi or single $MnSi_{1.7}$ phases are formed at high and low annealing temperature, respectively. The skyrmion window is proved to be enhanced in both temperature range or magnetic field range, which is useful for future applications and the investigation of magnetic skyrmion dynamics.

In the fifth chapter I discuss the Curie temperature of MnSi thin films. We have grown MnSi films with a broad variation regarding their structural properties and do not find a correlation between the Curie temperature and the structural variants. The problem remains unsolved. At the end, I present a short summary and some suggestions for the future work.

2. Experiment

2.1 Sample preparation

2.1.1 DC magnetron sputtering

Direct current (DC) Sputtering [127] is one kind of thin film physical vapour deposition (PVD) coating technique [128, 129]. A target material to be used as the coating is bombarded with ionized gas molecules causing atoms to be "sputtered" off into the plasma. These vaporized atoms are deposited when they condense as a thin film on the substrate. DC magnetron sputtering is the most powerful and inexpensive type of sputtering for metal deposition for electrically conductive target coating materials. There are two major advantages of using DC as a power source for this process: it is easy to control and it is a low cost option if one is doing metal deposition for coating.

DC magnetron sputtering is used extensively in the semiconductor industry for creating microchip circuitry on the atomic size. It is also used for gold sputter coatings of jewellery, watches and other decorations, for non-reflective coatings on glass and optical components, electrodes, as well as for metalized packaging plastics. In the basic configuration of (if the target material to be used as coating) a DC magnetron sputtering coating system is placed in a vacuum chamber with target parallel to the substrate.

The vacuum chamber is evacuated to a base pressure for removing H_2O, Air. Then it is backfilled with a high purity inert protect gas. The protect gas is usually Argon due to its relative large mass and the ability to transport kinetic energy upon impact during high energy molecular collisions in the plasma. In the sputtering process, plasma from Ar gas ions is the primary driving force of sputter thin film deposition. Typical sputter pressures range from 0.5 mTorr to 100 mTorr. Then a DC voltage typically in the -2 to -5 kV range is applied to the target coating material that is the cathode or the position at which electrons enter the system known as the negative bias. A positive charge is also applied to the substrate to be coated (or in other words, the anode). The electrically neutral Ar gas atoms are first ionized as a result of the forceful collision of these gas atoms onto the surface of the negatively charged target. The target will eject atoms off into the plasma – a hot gas-like state consisting of roughly half gas ions and half electrons.

The ionized Ar gas atoms are then driven to the substrate, which is the anode or positively biased attracting ionized gas ions, electrons and the vaporized target coating atoms. Those vaporized target atoms condense and form a thin film coating on the substrate to be coated, as shown in Figure 2.1 (e).

Though DC magnetron sputtering is on economical solution for different metal coatings, it has limitations to deposit ferromagnetic films and non-conducting dielectric insulating materials. Compared with traditional nonmagnetic targets, the high magnetic susceptibility of ferromagnetic targets like, Fe, Co, Ni, has electromagnetic shielding effects, thus that the magnet cannot affect the movement of electrons. Another problem of ferromagnetic targets is magnetic pinching of the plasma. There are different methods to solve the problem of depositing ferromagnetic films by sputtering: (1) Target design and improvement; (2) Enhanced magnetic field source for magnetron sputtering cathodes; (3) Reduce the magnetic permeability of the target; (4) Design a new magnetron sputtering system; (5) Design a new sputtering cathode device.

Non-conducting dielectric insulating materials accumulate charge, which can result in some quality issues like arcing, or the poisoning. The target material with a charge can result in the complete cessation of sputtering. To overcome these limitations of DC magnetron sputtering, several more complicated technologies have been developed such as Radio Frequency (RF) Sputtering, and High Power Impulse Magnetron Sputtering (HIPIMS) [130, 131]. RF sputtering alternates the electrical charge at a radio frequency to prevent a charge build up on the target or the coating material. HIPIMS utilizes a very high voltage, short duration burst of energy focused on the target coating material to generate a high density plasma. This can result in a much high degree of ionization of the coating material in the plasma. Therefore, HIPIMS usually has much higher deposition rate than the simple DC magnetron sputtering.

However, the power sources for RF or HIPMIS require much more complicated configuration, cabling and higher energy costs. DC magnetron sputtering is still the low-cost solution for many types of vacuum metal deposition like gold sputtering and other electrically conductive coatings.

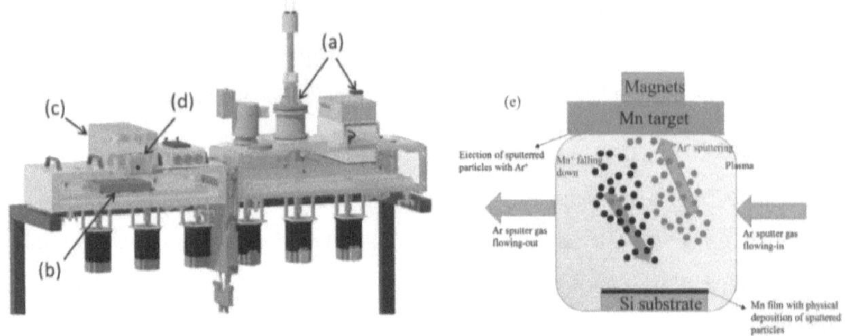

Figure 2.1. Schematic of DC magnetron sputtering combined with Flash Lamp Annealing g: (a) target and current controller, (b) substrate carrier, (c) flash lamp annealing chamber, (d) pre-heating system; (e) Schematic of DC magnetron sputtering process. Reproduced from [132].

The schematic of our DC magnetron sputtering is shown in Figure 2.1. We combine flash lamp annealing with DC magnetron sputtering to minimize the oxidation. Preparation of transition metals silicide films is our research object and I focus on MnSi. The transition metals oxidize very quickly, especially for the Mn element. Other transition metals like Fe and Co can form a thin oxide layer to prevent the further oxidizing. However, Mn will oxidize until it runs out. So the Mn target should be stored carefully. The deposition and annealing process must be performed in high vacuum condition without breaking the vacuum. Before depositing the Mn onto the Si substrate, the native SiO_2 from the Si substrate is removed by dipping into HF for 5 seconds. After that, the Si substrate is cleaned by acetone and alcohol. Then it is put into the vacuum chamber of sputtering as soon as possible. As shown in Figure 2.1, our sputtering chamber includes two targets and a current controller at (a), which can adjust the deposition rate. Part (b) is the sample carrier, which can be moved by 6 motors below. Part (c) is the flash lamp chamber, which can in-situ anneal the deposited film. Part (d) is the pre-heating system, which can reach a temperature of around 300 °C to remove H_2O attached to the substrates. The Mn target is installed at the right of this device, shown in (a) of Figure 2.1. Before sputtering, the Mn target will be pre-sputtered for 5 min to remove the oxide surface. Then the Si substrate on the sample carrier (b) will be moved in a reciprocating manner in the chamber to make sure the homogeneous deposition. The deposition rate was calibrated and we have deposited 7, 15, 20, 30, 40, 60 nm Mn film on Si substrates by this device.

2.1.2 Sub-second annealing

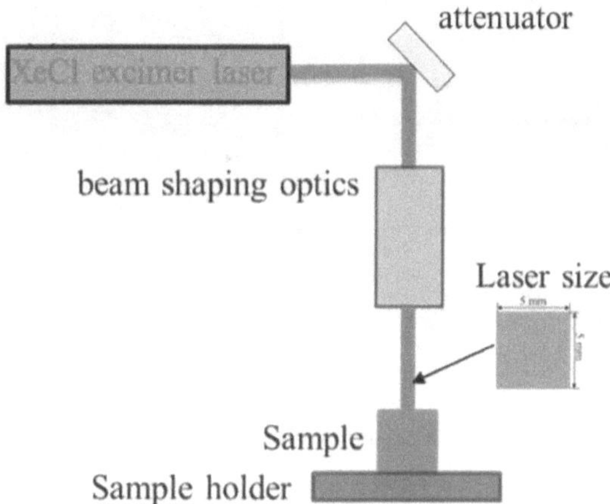

Figure 2.2 Schematic of pulsed laser melting: the laser has a wavelength of 308 nm and a pulsed duration of 30 ns.

In our group, we have two kinds of sub-second annealing methods. One is pulsed laser melting, which allows for a liquid epitaxy growth and another one is flash lamp annealing, a solid epitaxy growth method. Figure 2.2 shows the schematic of pulsed laser melting (PLM) [133]. The laser is generated by XeCl excimer gas and the energy can be adjusted by the attenuator or by adjusting the angle of two glasses, which controls the intensity of incident laser. The beam then passes through the beam shaping optics and a homogenized beam of 5 mm x 5 mm is obtained and can be applied to the surface of sample. PLM is a fast annealing on nanosecond time scale, which is a promising tool to achieve high doping concentration for semiconductor materials by liquid-phase epitaxy. The surface layer can be molten and recrystallized in nanoseconds. Dopant segregation, surface evaporation, and cellular breakdown always happen in the traditional annealing methods due to the long annealing time. However, this nanosecond PLM can inhibit these problems mentioned above. We try to use this method to form B20 MnSi films. The operation was at atmosphere. Thus O can react with our samples. We have tried to anneal our films with energy density of 50, 100, 150, 200 and 250 J/cm^3. Unfortunately, this method did not succeed in fabricating B20 MnSi films, while higher manganese silicides and Mn oxides can be formed.

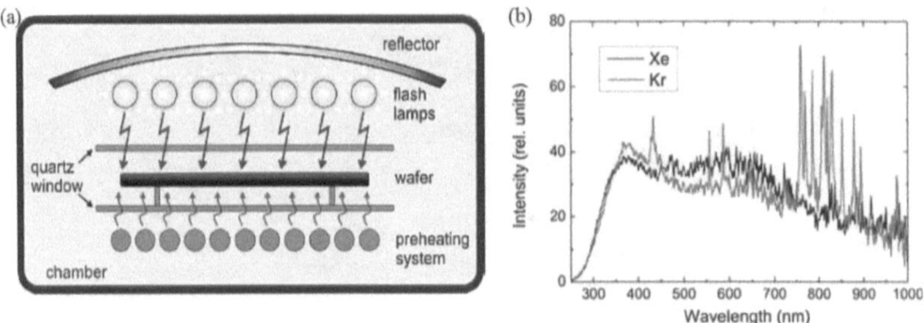

Figure 2.3. (a) Schematic of flash lamp annealing. (b) Time-integrated spectrum of a 3 ms flash from lamps filled with Xe or Kr at a pressure of 575 Torr. Reproduced from [124].

Figure 2.3 (a) shows the schematic of flash lamp annealing (FLA). FLA is a fast annealing on millisecond time scale, which is also a promising tool to achieve high doping concentrations semiconductor materials by solid-phase epitaxy. The FLA system consists of a reflector, a bank of Xe flash lamps, a wafer holder, two pieces of quartz plate and a preheating module. The 12 side-by-side connected Xe lamps with 30 cm can provide a large area of uniform flash. That makes FLA suitable for large-scale processing, up to 4 inch wafer. The operation process of FLA can be performed at N_2 or other gas flow, which can protect the sample from oxidizing continuously. The flash lamp and the processing chamber are separated, because the lamps should not be affected by sublimated or evaporated materials, and samples can be protected even if the bulb explodes or breaks. The reflector is designed to obtain the uniform temperature of the wafer. The preheating system consisting of a continuous halogen array that is installed under the sample processing chamber to increase the temperature of sample for high temperature reaction. When the temperature gradient between the front and backside is reduced, the preheating system enables high peak temperatures. This temperature gradient reduction prevents the accumulation of harmful stress levels in the substrate due to different thermal expansion of the front and backside. The sample processing chamber consists of two sheets of quartz panels for near ultraviolet and visible light, which separates the air and the protective atmosphere filled inside. The water-cooled annealing chamber can introduce different gases and the annealing can be carried out under different ambient gases. Figure 2.3 (b) displays the time-integrated spectrum of a 3 ms flash from lamps filled by Xe or Kr at a pressure of 575 Torr.

The pulsed energy of FLA is adjusted by voltage, which can charge and discharge the capacitor. The flash pulse duration depends on the discharge R-C circuit. The voltage applied to the parallelly connected flash lamps decays exponentially. The formula is as follows:

$$V(t) = V_0 \cdot e^{-t/RC} \tag{11}$$

where V_0 is the capacitor voltage at t = 0, R is the resistance of the discharge circuit, C is the capacitance of the discharge circuit. The discharge time can be extended by enhancing the resistance and / or capacitance in the circuit. The pulse duration range of our FLA is from 100 µs to 100 ms. I used the pulse durations of 3 ms, 6 ms and 20 ms in this book. The energy fluences of FLA for 3 ms and 20 ms flash are about 100 J/cm^2 and 250 J/cm^2, respectively. By charging and discharging the capacitor, the flash lamps can produce high density light. The sample can absorb and reflect the light. The temperature of samples is increased by absorbing this light. The absorbed energy density A can be calculated by the equation:

$$A = (1 - R_S)\Gamma E_D \tag{12}$$

Where E_D is the incoming energy density, R_S is the average reflectance of the sample and Γ is the correction for chamber wall reflections. The incoming energy density can be regulated by the voltage. The corresponding equilibrium temperature T_{equ} of the surface can be calculated by the following equation:

$$T_{equ} \approx T_0 + \frac{A}{C_p \rho d} \tag{13}$$

where T_0 is the starting temperature (either room temperature or a certain preheating temperature), C_p is the specific heat capacity of the sample, d is the sample thickness, and ρ is the mass density. This approximation assumes that the temperature dependence of C_p is negligible and that heat dissipation by radiation or convection takes place on a time scale much longer than the pulse length. The final temperature of the annealed surface layer could reach more than 2000 K depending on the intensity of the light flash and the absorbance of the annealed material.

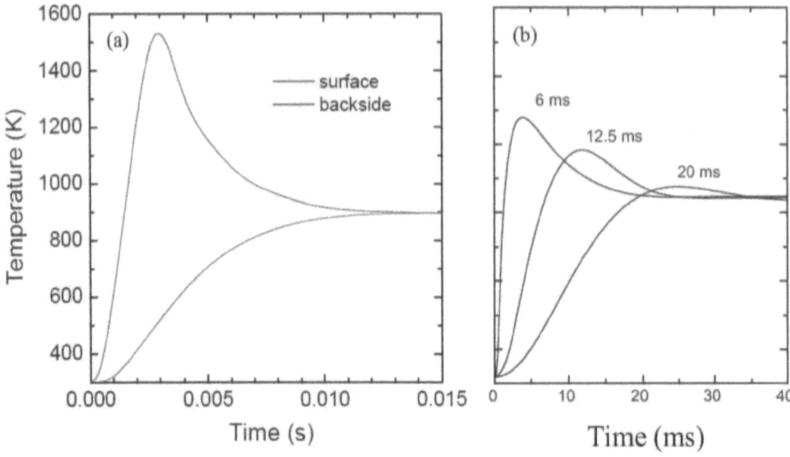

Figure 2.4. (a) The surface and backside annealing temperatures dependent on time with 3 ms pulse flash and 100 J cm^{-2} (b) the surface temperature dependent on time after pulses with different durations and 60 J cm^{-2}. Reproduced from [124, 134].

The time-dependent temperatures for a pure silicon wafer with 275 µm thickness from surface and backside are shown in Figure 2.4 (a). Two sides reach the same equilibrium temperature after around 13 ms. The time-dependent surface-temperature after flash pulse with the pulse duration of 6 ms, 12.5 ms and 20 ms is given in Figure 2.4 (b). If the energy density is the same, with a short pulse, the surface temperature goes higher than with a long pulse and the equilibrium temperature stays constant after around 40 ms [1]. Some thermal induced negative effects (oxidation, decomposition and defect generation) can be minimized by FLA. In this book, I mainly used FLA as the annealing tool.

As shown in Figure 2.5 for Mn films deposited on Si, depending on the annealing temperature, two reactions can happen for our samples at the same time [133]. In this process, diffusion and nucleation happen at the same time. Because the size of Si atoms is smaller than Mn, Si diffuses into Mn much more easily. The diffusion of Si dominates the phase formation process. One reaction is that Mn and Si atoms react and form MnSi$_{1.7}$ directly, as shown by reaction equation (14) and Figure 2.5. Another reaction can form B20 MnSi at the beginning, as shown by reaction equation (15). However, due to the Si wafer can provide endless Si atoms, MnSi can react with the diffused Si atom, forming Si rich phase, namely MnSi$_{1.7}$, as shown by reaction equation (16). In this recrystallization process, MnSi$_{1.7}$ is more easily formed than

MnSi. So most of MnSi thin films contain $MnSi_{1.7}$ impurities. This is also true for MnSi films grown by other solid state reaction or molecular beam epitaxy. As known from the mechanism of fast annealing on phase separation, this method can promote the reaction equation (15) and inhibit the reaction equation (14) and (16) by controlling the different annealing temperatures with ultra-fast heating rates. The sample was annealed from the surface and can reach over 2000 K in some milliseonds. The fast non-equilibrium process attains the temperature, where B20 MnSi can nucleate in milliseconds. Moreover, by properly selecting flash power, the nucleation of the extra $MnSi_{1.7}$ phase can be successfully suppressed due to the high heating rate.

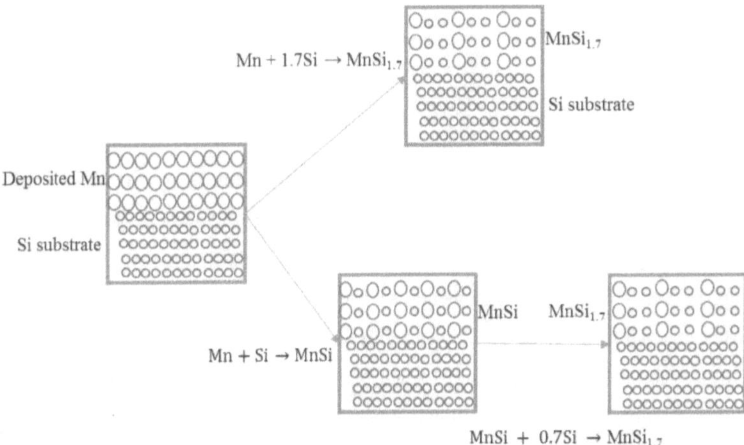

Figure 2.5. Schematic of atom diffusion and reaction. The red circle and blue circle stand the Si and Mn atom, respectively. The upper picture is the reaction (14) for forming MnSi$_{1.7}$ directly. The down left shows the reaction (15) for forming B20 MnSi directly. The down right picture shows the reaction (6) that MnSi reacts with extra Si forming MnSi$_{1.7}$.

$$Mn + 1.7Si \rightarrow MnSi_{1.7} \tag{14}$$

$$Mn + Si \rightarrow MnSi \tag{15}$$

$$MnSi + 0.7Si \rightarrow MnSi_{1.7} \tag{16}$$

2.2 Structure characterization: X-ray diffraction

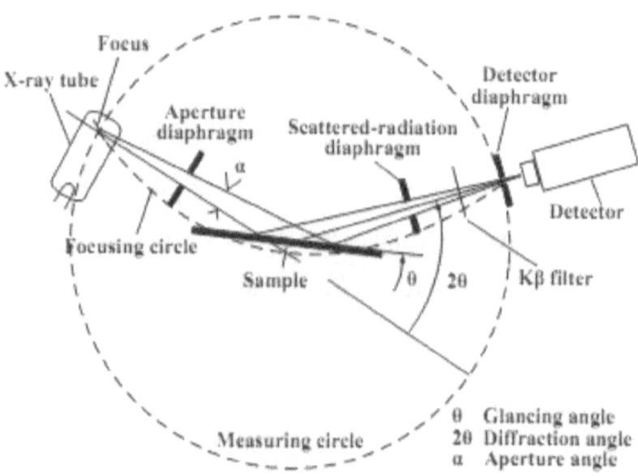

Figure 2.6. Schematic of X-ray diffraction. Reproduced from [135].

To analyse the crystal phases of the FLA-treated Mn/Si film, X-ray diffraction (XRD) was measured. XRD was performed by a Bruker D8 Advance diffractometer. Bragg-Brentano-geometry with a graphite secondary monochromator and a scintillator was used in this Bruker D8 setup [1]. Small angle grazing incidence XRD measurements were conducted using a Smart Lab diffractometer from Rigaku, equipped with a rotating Cu-anode operated at 9 kW. XRD in the θ-2θ geometry and grazing-incidence XRD (GIXRD) was performed to probe the crystallization and crystallite size and we always use a parallel beam. For the sample grown on Si(100) substrate, the MnSi film is polycrystalline and the XRD peaks measured by D8 are very weak. Thus the grazing-incidence XRD was applied on these samples. In contrary, the samples on Si(111) substrate are always textured. Thus D8 can detect the crystal products. For the MnSi epitaxy film, we also measured the azimuthal-angle scans (phi scan) and pole figure to determine the crystal symmetry.

The schematic of X-ray diffraction is shown in Figure 2.6. X-ray diffractometers consist of three basic elements: an X-ray tube, a sample holder, and an X-ray detector. X-rays are generated in a cathode ray tube by heating a filament to produce electrons, accelerating the electrons toward a target by applying a voltage, and bombarding the target material with electrons. When electrons have sufficient energy to dislodge inner shell electrons of the target material, characteristic X-ray spectra are produced. These spectra consist of several components. The most common components are K_α and K_β. K_α consists, in part, of $K_{\alpha 1}$ and $K_{\alpha 2}$. Generally, $K_{\alpha 1}$ has a slightly shorter wavelength and twice the intensity as $K_{\alpha 2}$. The specific

40

wavelengths are characteristic of the target material (Cu, Fe, Mo, Cr). Filtering, by foils or crystal monochromators, is required to produce monochromatic X-rays needed for diffraction. $K_{\alpha 1}$ and $K_{\alpha 2}$ are sufficiently close in wavelength such that a weighted average of the two is used. Copper is the most common target material for single-crystal diffraction, with Cu K_{α} wavelength = 1.5418 Å. These X-rays are collimated and directed onto the sample. For θ-2θ scans, as the sample or the X-ray source and the detector are rotated, the intensity of the reflected X-rays is recorded. When the geometry of the incident X-rays impinging the sample satisfies the Bragg Equation, constructive interference occurs and a peak in intensity occurs. A detector records and processes this X-ray signal and converts the signal to a count rate as the output signal. The geometry of an X-ray diffractometer is such that the sample rotates (or in relative) in the path of the collimated X-ray beam at an angle θ while the X-ray detector is mounted on an arm to collect the diffracted X-rays and rotates at an angle of 2θ. To enhance the detection sensitivity for thin film samples, small angle grazing incident XRD can be used. For that, X-rays enter the sample at a small grazing angle and can have interact with a larger volume of the detected film. The incident angle is fixed and only the detector (2θ is moving).

2.3 Property characterization

2.3.1 Magnetic properties

In this book, all the magnetic properties are measured by a commercial vibrating sample magnetometer equipped with a superconducting quantum interference device (SQUID-VSM). First the liquid helium consumption is about 5 L per day which is much less than for the old generation SQUID-MPMS (magnetic property measurement system) due to the special design and the protection by liquid nitrogen. Second, less than 15 mins are required to bring the temperature from 300 K down to 5 K since the heating and cooling rate can reach 35 K/s. This is also much faster than SQUID-MPMS. Third, the measurement speed is 5-6 times faster with the same sensitivity due to the advanced detection system. The setup is shown in Figure 2.10(a).

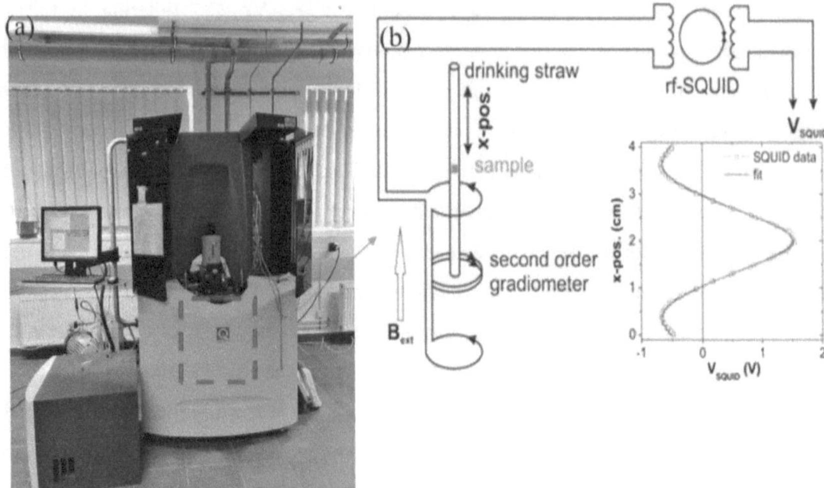

Figure 2.10. Superconducting quantum interference device Vibrating Sample Magnetometer (a) and schematic for the signal pickup (b). Reproduced from [136].

Figure 2.10(b) illustrates a simplified process of the detection for SQUID-VSM. The superconducting coil can produce an external magnetic field B_{ext} up to 7 T with a direction parallel to the sample moving direction. The superconducting detection coils are composed of a second-order gradiometer. This superconducting detection coils usually detects the change of magnetic flux created by mechanically moving the sample through a superconducting pick-up coil which is converted to a voltage. Only an induced current by a disturbance to the corresponding uniform magnetic field will be detected. That is related to the sample. Then, the current is coupled to SQUID for measurement. The SQUID converts the current to voltage with high sensitivity. This device is sensitive to 1×10^{-8} emu. If the signal is too weak, there will be some errors arising from the sample geometry or sample holder.

For measuring the sample, several procedures should be done. First for sample mounting, a quartz sample holder is used for bulk or thin film samples and a brass sample holder is used or powder samples loaded into a capsule. Since the magnetic field is parallel to the sample moving direction (perpendicular to the ground), my thin film samples can be glued to the quartz sample holder with a parallel or perpendicular direction, namely for in-plane or out-of-plane measurements, respectively. For the in-plane measurement, the sample size should be smaller than 5×5 mm^2. For the out-of-plane measurement, the sample size should be smaller than 2.5×5

mm^2. The position of the sample on quartz should be 66±3 mm from the end of the sample holder. The sample holders and the sample position for the VSM are shown in Figure 2.11(a).

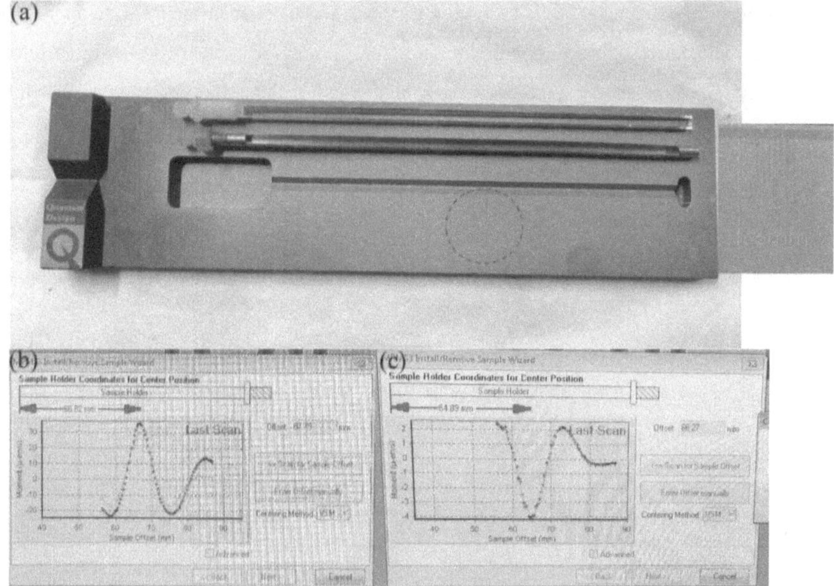

Figure 2.11. (a) Sample holders and the sample position for SQUID-VSM measurements, (b) a center scan for a ferromagnetic sample and (c) a center scan for a diamagnetic sample.

Afterwards, the sample should be centred by performing measurements along the length to get the sample position. To get the real position, the cenetring process is always repeated three times with the vibration amplitude of 1, 2 and 4 mm to get the real position. From the cenetring process, we can get two informations. First, if there is a positive peak shown in Figure 2.11(b), it means this material is paramagnetic or ferromagnetic. If there is a negative valley as shown in Figure 2.11(c), it indicates the material is diamagnetic or the measurement temperature is far above the Curie temperature. Second, from the moment value, one can calculate the magnetic susceptibility roughly. After centring, one can start to measure the sample by entering the measurement sequence. The sequence can include the variation of magnetic field and temperature. The details are included in the subchapters of this book.

2.3.2 Magneto-transport properties

The magnetic field- and temperature-dependent electrical properties of a material reflect its electronic and magnetic structure. In my book, we have measured the temperature-dependent magnetoresistance to obtain the Curie temperature. The magnetic-field dependent magnetoresistance is also measured to get the spin texture transition. It is known that B20 MnSi can host Skyrmions, which produces a topological Hall effect.

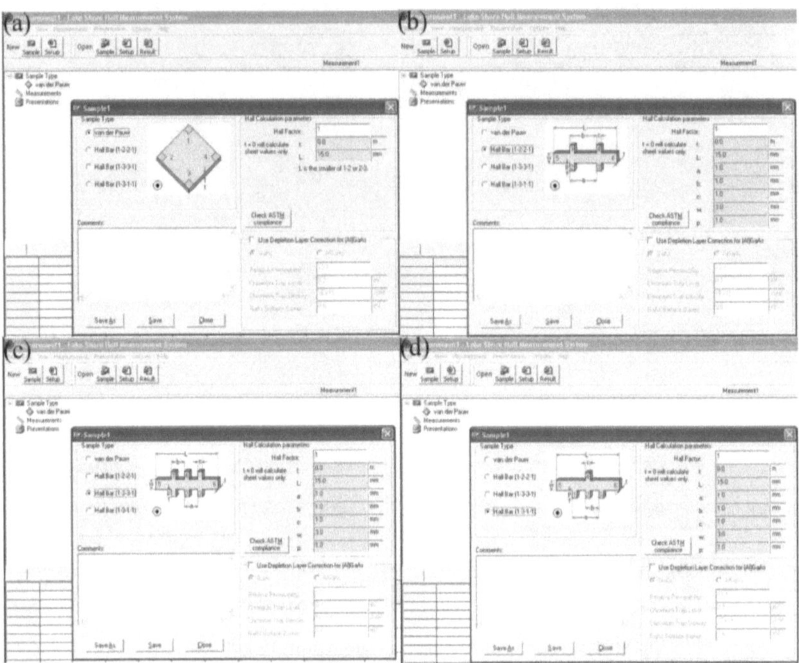

Figure 2.12 (a) Van der Pauw geometry and (b-d) Hall bar geometry.

Here, a fully integrated Hall measurement system (HMS) Model 9700A from Lake Shore is used to characterize electrical properties, i.e. magnetoresistance, Hall resistance, and temperature dependent sheet resistance with or without magnetic field. For all electrical measurements, a sample resistance ranging from 0.04 mΩ to 200 GΩ is within the detection limits of our setup. When measuring the magnetoresistance and Hall effect, a superconducting magnet is employed to set the magnetic field up to $\pm$7 T, and the temperature can be varied from 1.8 to 400 K by liquid Helium cooling. There are four types geometry for the Hall measurements in our setup. One is Van der Pauw method and three are Hall bar geometries, shown in Figure 2.12. For my materials, I always use Van der Pauw method to measure the resistance and Hall Effect.

After flash lamp annealing, we check the crystalline quality of the regrown layers by Raman and XRD. Then we select the samples for the magneto-transport measurements. Before measurement, we need to prepare four electrodes for the 5×5 mm^2 sample at the corners for to the Van der Pauw geometry. Either Indium or silver pastes connect the silver wire to the sample. Tin welding can connect the other end of the silver wire to the metal cylinder of the sample holder. Since my samples have good conductivity compared with semiconductor materials, we can easily get good Ohmic contacts. For magnetic materials, from the resistance dependent temperature curves, one can calculate dR/dT, which has one peak indicating the magnetic transition temperature. From the temperature dependent magnetoresistance, we also can find there is a valley around the Curie temperature. Below the Curie temperature, it shows negative magnetoresistance due to the ferromagnetic nature of my samples.

3.. B20-MnSi films grown on Si(100) substrates with magnetic skyrmion signature

Magnetic skyrmions have been suggested as information carriers for future spintronic devices. As the first material with experimentally confirmed skyrmions, B20-type MnSi has the research focus for decades. Although B20-MnSi films have been successfully grown on Si(111) substrates, there is no report about B20-MnSi films on Si(100) substrates, which would be more preferred for practical applications. In this chapter, we present the first preparation of B20-MnSi on Si(100) substrates. It is realized by sub-second solid-state reaction between Mn and Si via flash-lamp annealing at ambient pressure. The regrown layer shows an enhanced Curie temperature of 43 K compared with bulk B20-MnSi. The magnetic skyrmion signature is proved in our films by magnetic and transport measurements. The millisecond-range flash annealing provides a promising avenue for the fabrication of Si-based skyrmionic devices.

The results presented here are under preparation for publication: Z. C. Li, Y. Yuan, R. Hübner, V, Begeza, L. Rebohle, M. Helm, K. Nielsch, S, Prucnal S. Q. Zhou, B20-MnSi films grown on Si(100) substrates with magnetic Skyrmion signature. The TEM measurements were done by R. Hübner. All other measurements were done by the book author. The book author also analyzed all data and wrote the manuscript.

3.1 Introduction

In noncentrosymmetric magnetic materials, the Dzyaloshinskii Moriya (DM) interaction can cause spins to arrange non-collinearly [99, 137, 138] and together with the ferromagnetic exchange interaction, it results in a whirling domain structure called skyrmions [44, 139-142]. Due to the small size and smaller motion energy for skyrmions, they can be used in new generation integrated storage devices [40, 143]. Magnetic skyrmions were experimentally observed for the first time in bulk ferromagnetic B20-type MnSi [5]. Afterwards, a lot of effort has been devoted to the growth of B20-MnSi thin films [72, 100, 101, 142]. Due to the small lattice mismatch, the films were mostly grown on Si(111) substrates. For instance, B20-MnSi films on Si(111) were successfully prepared by molecular beam epitaxy (MBE) [100, 101, 144], by solid-state phase epitaxy (SPE) [72, 102], and magnetron sputtering [145]. In most of the cases, MnSi films grown on Si(111) show a Curie temperature of around 43 K, i.e. much higher than the 29.5 K for bulk MnSi. However, so far, there has been no report about the formation of B20-MnSi on Si(100) substrate, which is more preferred for spintronic devices due its compatibility with complementary metal-oxide-semiconductor (CMOS) technology [146]. It seems that the growth of B20-MnSi on Si(100) is not possible due to the large mismatch of 16% [147]. Indeed, the $MnSi_{1.7}$ phase is more easily grown on Si(100) substrates and captures researchers' attention due to its excellent thermoelectric properties [148, 149]. $MnSi_{1.7}$ films were prepared on Si(100) by SPE [150] and MBE [151]. Furthermore, Wu *et al.* and Kahwaji *et al.* found that B2-MnSi can be formed on Si(100) substrates [152, 153]. However, B2 MnSi can not host skyrmions since it is centrosymmetric [153]. The absence of B20-MnSi on Si(100) substrates significantly limits the application of skyrmions in integrated spintronics devices. Therefore, it is desire to search for different growth approaches for B20-MnSi on Si(100) substrates. Xie et al. have demonstrated the growth of Mn_5Ge_3 films on Ge(100) by millisecond solid-state reaction between Mn and Ge via flash-lamp annealing [154]. Note that, by normal solid-state reaction Mn_5Ge_3 is more easily formed on Ge(111) substrates [155]. This therefore suggests that a non-equilibrium annealing method could allow for the nucleation of a crystalline phase, which is otherwise not favourable under thermal equilibrium condition.

In this chapter, we use millisecond flash-lamp annealing induced solid-phase reaction to synthesize B20-MnSi films. We show that the nucleation of B20-MnSi on Si(100) substrates is possible, leading to a continuous polycrystalline thin film, although the $MnSi_{1.7}$ parasitic phase cannot be eliminated. The Curie temperature of the fabricated films, compared to bulk B20-MnSi, increases to around 43 K. The characteristic signature of magnetic skyrmions in our MnSi thin films is proved by magnetization and magneto-transport measurements [101]. The skyrmions can be stabilized from 1000 Oe to 7000 Oe at the whole temperature range below 43 K.

3.2 Experiment

To grow B20-MnSi films, we first deposited Mn films with thicknesses from 7 nm to 30 nm on Si(100) substrates by magnetron sputtering. Afterwards, millisecond flash-lamp annealing (FLA) with 20 ms pulse duration in a continuous N_2 flow was used to heat the sample and to trigger the solid-phase reaction between Mn and Si [124]. The thickness of the formed MnSi films is from around 14 nm to 60 nm, i.e. around the twice of the deposited Mn films as checked by transmission electron microscopy. Such a non-equilibrium process leads to a very high phase formation speed, which makes it possible to realize non-equilibrium materials phases, like B20-MnSi on Si(100), which is otherwise not possible under thermal equilibrium. The topography of the regrown layers was observed by visible-light microscopy. To analyze the microstructure of the FLA-treated MnSi films, micro-Raman spectroscopy and transmission electron microscopy (TEM) were employed. The micro-Raman experiments were done using a Horiba micro-Raman system with the excitation wavelength of 532 nm, and the signal was recorded with a liquid-nitrogen-cooled silicon CCD camera. Bright-field and high-resolution TEM images were recorded with an image-C_s-corrected Titan 80-300 microscope (FEI) operated at an accelerating voltage of 300 kV. High-angle annular dark-field scanning transmission electron microscopy (HAADF-STEM) imaging and spectrum imaging analysis based on energy-dispersive X-ray spectroscopy (EDXS) were performed with a Talos F200X microscope (FEI) operated at 200 kV to obtain the element composition. The magnetic properties of the films were measured by a superconducting quantum interference device equipped with a vibrating sample magnetometer (SQUID-VSM) with the field parallel (in-plane) to the films. The transport properties of MnSi films were investigated by a Lake Shore Hall measurement system. Magnetic-field (in-plane) dependent resistance was measured between 5 and 45 K using the van der Pauw geometry.

3.3 Results and Discussions

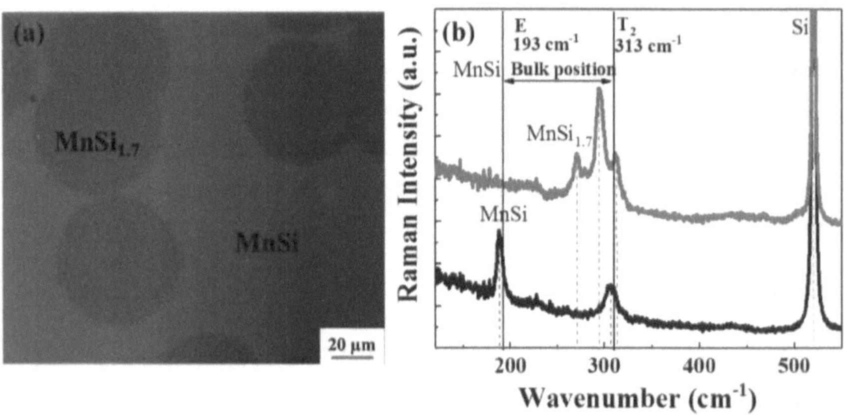

Figure 3.1. (a) Visible-light microscopy image of the sample annealed from a 30-nm-thick Mn layer on Si(100) substrates. The grey concentric circles are the phase of MnSi$_{1.7}$ and the featureless area is B20-MnSi. The scale bar is 20 μm. (b) Raman spectra obtained at different positions as shown in Figure 3.1 (a). For the darker concentric circles and the matrix in-

between, the Raman spectra were measured for more than 6 positions. Each matrix position shows the same Raman signal as the black curve, pointing to the B20-MnSi phase. The spectrum of each circular area shows three peaks at around 300 cm^{-1}, consistent with the MnSi$_{1.7}$ phase.

We first focus on the sample grown from 30 nm Mn on Si(100). A representative visible-light microscopy image of the regrown layer on Si(100) is shown in Figure 3.1 (a). In this image, there are two different areas: one is composed of darker concentric circles with a diameter of around 60 μm and the other one is the matrix area in-between. To understand the phase formation for these two areas, the corresponding room-temperature Raman spectra are shown in Figure 3.1 (b). All presented Raman spectra show the peak at around 520.5 cm^{-1} (dashed green lines), which corresponds to the transverse/longitudinal optical (TO/LO) phonon mode in the Si substrate [156]. Additionally, the matrix area shows two well-separated peaks at about 188 and 307 cm^{-1} (dashed black lines), which correspond to the T2 Raman-active phonon modes in MnSi [157, 158]. In fully relaxed bulk MnSi, these two phonon modes are located at around 194 and 316 cm^{-1}. The shift of the phonon mode positions towards lower wavenumbers indicates the existence of in-plane biaxial tensile strain. The Raman spectrum obtained from the area of the concentric circles exhibits three separated peaks at around 300±20 cm^{-1} (dashed red lines). These peaks confirm the formation of the MnSi$_{1.7}$ phase. These three phonon modes should account for the slightly different neighbourhoods of the Mn–Si bonds in MnSi$_{1.7}$ [159]. Based on the Raman results, we can conclude that the circular and matrix areas exhibit the MnSi$_{1.7}$ and B20-MnSi phase, respectively. More specifically, MnSi$_{1.7}$ is embedded in the B20-MnSi matrix. By rough estimation from the visible-light microscopy images, the areal fraction of the MnSi phase comprises around 45%. A more reliable estimation is given by comparing the saturation magnetization of our films with that of bulk MnSi in a later session. The precipitation of MnSi$_{1.7}$ in the shape of concentric circles is assumed to be caused by an inhomogeneous temperature distribution or compositional fluctuations [160]. We notice that the formation of MnSi$_{1.7}$ within the sample of around 10×10 mm^2 is random. Inevitably, some local regions have lower temperature or redundant Si, both favouring the nucleation of MnSi$_{1.7}$. However, more investigations are needed for a complete understanding.

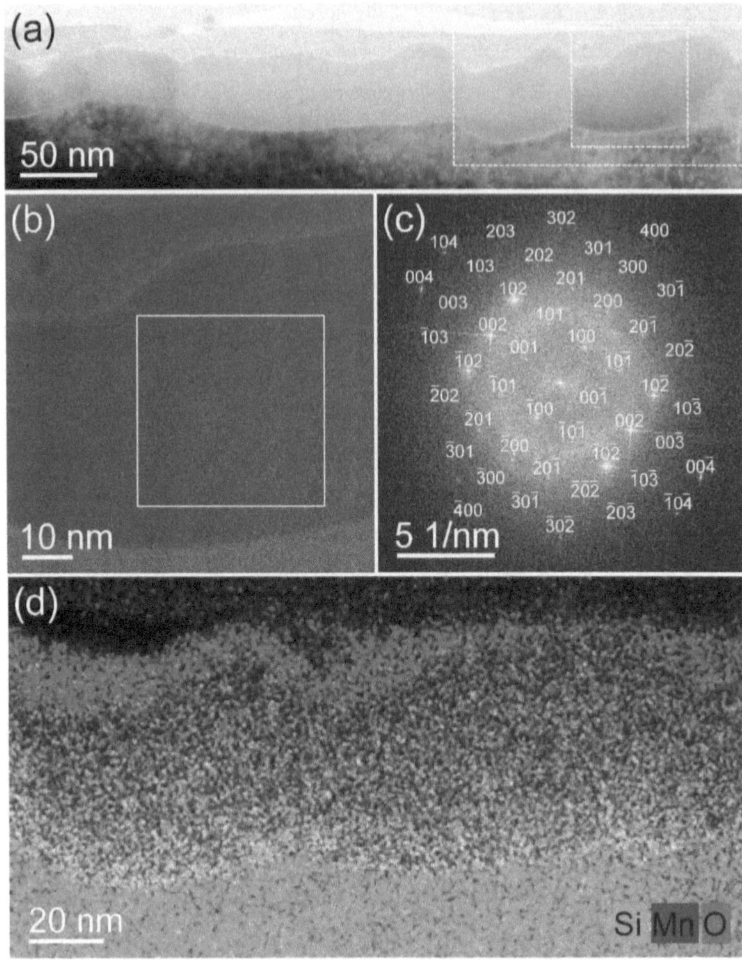

Figure 3.2. (a) Cross-sectional bright-field TEM image of the FLA-treated 30-nm-thick Mn layer on Si(100) substrate obtained from the matrix area shown in Figure 1(a). (b) High-resolution TEM image of the area marked with a dashed white square in panel (a). (c) Fast Fourier transform of the region marked with a white square in panel (b) and indexed based on a MnSi [0 $\bar{1}$ 0] zone axis pattern. (d) Superimposed EDXS-based element distributions (blue: manganese, green: silicon, red: oxygen) for the area marked with a white dashed rectangle in panel (a), confirming the presence of a continuous B20-MnSi film.

Figure 3.2 (a) shows a representative cross-sectional bright-field TEM image of the FLA-treated 30-nm-thick Mn layer on Si(100) obtained from the matrix area indicated in Figure 3.1

(a). Below a 5-to-20-nm-thin amorphous surface film (light-grey with uniform brightness), there is an about 50-nm-thick polycrystalline film which is mainly composed of grains having a height equal to and a lateral extension larger than the film thickness (dark-gray appearance with slightly varying orientation contrast). To further characterize the phase structure of the continuous polycrystalline layer, high-resolution TEM imaging (Figure 3.2 (b)) combined with fast Fourier transform analysis (Figure 3.2 (c)) was performed. Taking the B20-type MnSi structure, the diffractogram in Figure 3.2 (c) can be described by a [0 $\bar{1}$ 0] zone axis pattern (in particular: normal to the growth plane [3 0 2] MnSi is parallel to [0 0 1] Si and in-plane [0 $\bar{1}$ 0] MnSi is parallel to [1 $\bar{1}$ 0] Si). The spatial arrangement of the elements Mn, Si, and O was characterized by EDXS-based analysis in scanning TEM mode and is shown in Figure 3.2 (d). In particular, Mn and Si are homogeneously distributed within the silicide layer, and the Mn:Si atomic ratio is determined to be close to 1:1. Above the MnSi layer, there is an oxide layer composed of Mn and Si. The small oxygen signal within the Mn silicide region is caused by TEM lamella side-wall oxidation during storage in air. For thicker TEM lamella positions, this O signal is even less pronounced. In summary, TEM analysis confirms the formation of a continuous polycrystalline B20-type MnSi film.

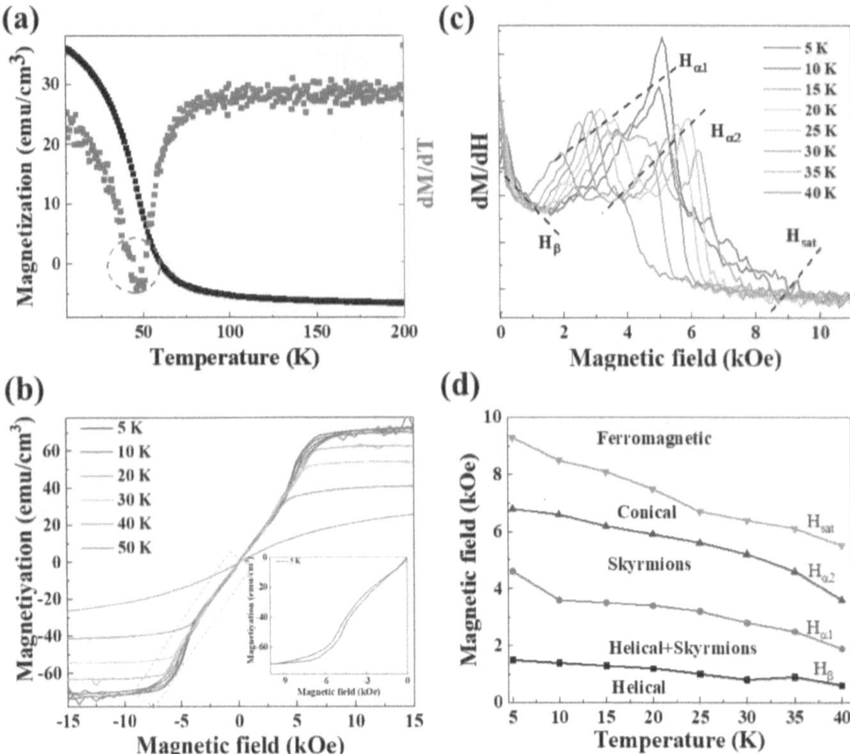

Figure 3.3. Magnetic properties of B20-MnSi made by solid-state reaction of a 30-nm-thick Mn layer on Si(100) substrate. (a) In-plane saturation magnetization (black squares) and the calculated derivative dM/dT (red squares). The valley of dM/dT at around 43 K indicates the Curie temperature. (b) In-plane M-H curves measured at different temperatures. (c) Static susceptibility in increasing field sweeps for the MnSi film at various temperatures. The black dashed lines in (c) show the shift of transition fields at each temperature. $H_{\alpha1}$ and $H_{\alpha2}$ bound the region of stable elliptic skyrmion. The dashed lines define the boundary for metastable helicoids H_β and ferromagnetic H_{sat}. (d) Magnetic phase diagram. With increasing temperature, the mentioned magnetic phases formed at lower magnetic fields.

Figure 3.3 (a) displays the temperature-dependent saturation magnetization of the MnSi alloy obtained from the 30-nm-thick Mn layer on Si(100) after FLA. As indicated by the minimum of the derivative dM/dT (red square) curve, the Curie temperature of our B20-MnSi film is 43 K and thus higher than that for bulk MnSi (29.5 K). In agreement with the Raman spectroscopy results, this Curie temperature enhancement is assumed to be caused by strain

[102]. It is known that $MnSi_{1.7}$ is a weak itinerant magnet with a saturation magnetization of 0.012 μ_B/Mn [82] compared with 0.43 μ_B/Mn for B20-MnSi [161], which is why the measured magnetization must originate from the B20-MnSi phase. The in-plane MH curves in Figure 3.3(b) show multi-hysteresis below the Curie temperature, which arises from the variety of different spin arrangements. To better observe the multi-hysteresis, the insert of Figure 3.3 (b) is an enlargement of the blue dashed rectangle. Our measurements also indicate that the transition between different magnetic structures occurs within the whole temperature range, not only near the Curie temperature as for bulk MnSi. The saturation magnetization at 5 K is around 72 emu/cm^3. If we neglect the magnetization of $MnSi_{1.7}$ and compare with the saturation magnetization of 163 emu/cm^3 for bulk MnSi, the MnSi volume fraction in our sample is around 45%. This value is in reasonable agreement with that obtained by calculating areal fraction from Figure 3.1 (a).

In Figure 3.3 (c), the derivatives of the static magnetization dM/dH show four critical transition fields, named as H_β, $H_{\alpha 1}$, $H_{\alpha 2}$, and H_{sat} with the dashed lines indicating their shift with temperature. Below H_β, the system stays at its ground helical state. The next apparent peaks at $H_{\alpha 1}$ and $H_{\alpha 2}$ mark the first-order transitions in and out of a new phase with a difference in topology, namely the so-called skyrmion phase [96]. According to this, the region between H_β and $H_{\alpha 1}$ is a transition region, where the ground helical and the skyrmion phase coexist. Above $H_{\alpha 2}$, the skyrmion phase almost vanishes and the system evolves into a conical state, followed by another transition to field-polarized ferromagnetism above the critical field of H_{sat}. Based on the temperature dependency of the transition fields a magnetic phase diagram as shown in Figure 3.3 (d), can be constructed. With increasing the magnetic field, the Helical, Helical + skyrmion, skyrmion, Conical and Ferromagnetism phase emerge in this order. It is interesting that our B20-MnSi film on Si(100) also hosts skyrmion over a wide temperature and magnetic-field range similar to what has been shown in literature [72, 117]. This enlarged skyrmion stability offers more flexibility in spintronic applications [162].

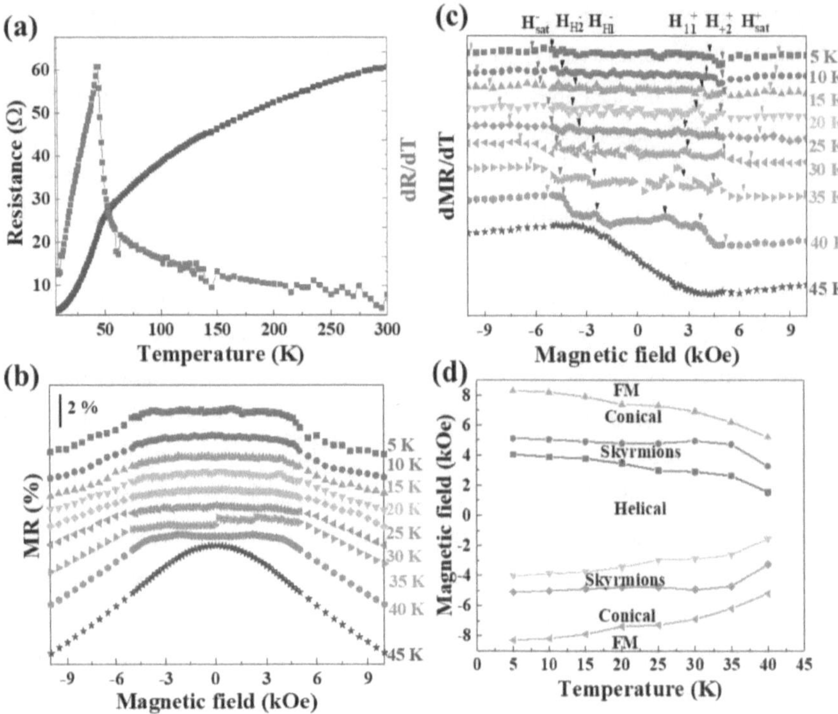

Figure 3.4. (a) Temperature-dependent resistance (black squares) and calculated dR/dT (red squares). The peak of dR/dT indicates the position of the Curie temperature. (b) Magnetic-field-dependent magnetoresistance at various temperatures. The anomalous plateau disappears at 45 K (above the Curie temperature). (c) Calculated derivative dR/dT from data in panel (b). The black, red, and green arrows stand for the position of $H_{\alpha 1}$, $H_{\alpha 2}$, and H_{sat}, respectively. (d) Magnetic phase diagram of the MnSi film with respect to temperature and magnetic field.

The in-plane electrical resistance of the MnSi film and the calculated dR/dT are shown as a function of temperature in Figure 3.4 (a). The increase of resistance with temperature clearly indicates the metallic behaviour of the MnSi film. The calculated dR/dT in Figure 3.4 (a) has a peak, indicating the Curie temperature around 43 K, in agreement with the magnetization measurement shown in Figure 3.3 (a). The magnetoresistance (MR), $MR = \frac{R_H - R_0}{R_0} \times 100\%$ (R_H: resistance under magnetic fields, R_0: resistance at zero field) of the MnSi film at various temperatures is shown in Figure 3.4 (b). There is an anomalous plateau at low magnetic fields and below the Curie temperature, which is due to the anomalous topology. Above the Curie temperature, this anomalous plateau disappears [163].

The field-driven evolution of the spin textures in MnSi can be investigated by the derivative dMR/dH from Figure 4 (b) [164]. As shown in Figure 3.4 (c), values for the positive and negative transition fields are denoted by the superscripts "+" and "−", respectively. With increasing magnetic field, the helimagnetic phase transforms into skyrmion via a first-order phase transition, manifested as peaks at $H_{\alpha 1}$ due to the completely different topological properties between helimagnetic and skyrmion phase. Above the critical field $H_{\alpha 2}$, the system transforms into the conical phase. Skyrmion can be stabilized in the range between $H_{\alpha 1}$ and $H_{\alpha 2}$. H_{sat} is the critical magnetic field, above which MnSi changes into the ferromagnetic state. To better observe the change of the transition fields, the magnetic phase diagram of B20-MnSi obtained by transport measurements is shown in Figure 3.4 (d). The magnetic skyrmions are stable in a broad field and temperature window, in agreement with the magnetization measurements.

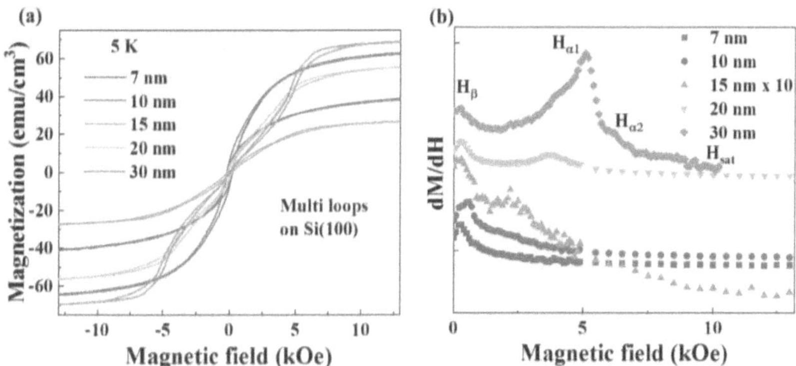

Figure 3.5. (a) In-plane MH curves (measured at 5 K) after FLA-treatment of Mn/Si(100) samples having different Mn layer thickness. The samples annealed from 20- and 30-nm-thick Mn show multi-hysteresis, indicating there is some magnetic phase transitions. (b) The calculated derivative dM/dH of the MH curves from (a). For the sample annealed from 7- and 10-nm-thick Mn, there is only one peak H_β, which is the transition from helical to concial state, since the thickness of the regrown layer MnSi is thinner than the theoretical size of MnSi Skyrmion. When the Mn layer is thicker than 15 nm, an extra $H_{\alpha 1}$ peak can be detected. To better check the change of the ample annealed from 15 nm Mn, the signal is magnified ten times.

To check the thickness dependence on the formation of B20-MnSi, Figure 3.5 (a) shows the MH curves obtained from MnSi layers made by solid-state reaction of Mn films with different thicknesses. Since the theoretical size of the skyrmion in MnSi is around 18 nm, the skyrmion can be detected only when the grain size of MnSi is larger than 18 nm. So, there is no multi-hysteresis, when the deposited Mn is too thin. With increasing the Mn thickness to 15 nm, first traces of the multi-hysteresis originating from the transitions of different magnetic structures can be observed, indicating the existence of magnetic skyrmion in the prepared B20-MnSi films [102]. The MnSi films prepared by ms-range solid-state reaction of 20- and 30-nm-thick Mn layers on Si(100) show obvious multi-hysteresis.

Figure 3.5 (b) shows the calculated derivatives dM/dH. All these curves have a peak H_β at around 300 Oe, which is the transition from the helical to skyrmonic phase. For samples with 7 and 10 nm Mn, the system changes from helical to conical due to the smaller size of the grains [165]. For thicker films, the neighbor phase is magnetic skyrmion. Since the signal of the sample annealed from 15 nm Mn is very weak compared with other samples, the dM/dH curve is multiplied by 10 to observe the change clearly. Moreover, for the sample annealed from 15, 20 and 30 nm Mn, there is also another peak $H_{\alpha 1}$, where the helical phase completely changes into the skyrmion phase. The peak $H_{\alpha 2}$ is the position, where the metastable skyrmion begin to change into the conical structure. Skyrmion are stabilized in the region from $H_{\alpha 1}$ to $H_{\alpha 2}$. The magnetic state finally changes into ferromagnetism at H_{sat}.

3.4 Conclusions

In summary, thin films with the B20-MnSi phase on Si(100) substrates are fabricated for the first time. The nucleation of B20-MnSi is believed to be triggered by the fast solid-state phase reaction between Mn and Si via ms-range flash-lamp annealing. Compared with the corresponding bulk material, the B20-MnSi films made by ms-range annealing show an increased Curie temperature of around 43 K. The magnetic and transport measurements reveal that Skyrmion in B20-MnSi on (100) Si made by sub-seconds solid-state reaction are stable within much broader field and temperature windows than bulk MnSi. The parasitic $MnSi_{1.7}$ phase can be further minimized or eliminated by optimizing the annealing conditions, the quality of the deposited Mn film, and its interface with the Si substrate. Our work demonstrates a promising option for the fabrication of B20-type transition metal silicides for integrated and/or hybrid spintronic applications by using Si(100) wafers, which are more preferable for industry applications.

4. Phase selection in Mn-Si alloys by fast solid-state reaction with enhanced skyrmion stability

B20-type transition-metal silicides or germanides are noncentrosymmetric materials hosting magnetic skyrmions, which are promising information carriers in spintronic devices. The prerequisite is the preparation of thin films on technology-relevant substrates with magnetic skyrmions stabilized at a broad temperature and magnetic-field working window. The canonical example is the B20-MnSi film grown on Si substrates. However, the as-yet unavoidable contamination with $MnSi_{1.7}$ occurs due to the lower nucleation temperature of this phase. In this chapter, we report a simple and efficient method to overcome this problem and prepare single-phase MnSi films on Si substrates. It is based on the millisecond reaction between metallic Mn and Si using flash lamp annealing (FLA). By controlling the FLA energy density, we can grow single-phase MnSi or $MnSi_{1.7}$ or their mixture at will. Compared with bulk MnSi the prepared MnSi films show an increased Curie temperature of up to 41 K. In particular, the magnetic skyrmions are stable over a much wider temperature and magnetic-field range than reported previously. Our results constitute a novel phase selection approach for alloys and can help enhance specific functional properties such as enhancing the stability of magnetic skyrmions.

The results presented here were published as: Z. C. Li, Y. F. Xie, Y. Yuan, Y. D. Ji, V, Begeza, L. Cao, R. Hübner, L. Rebohle, M. Helm, K. Nielsch, S, Prucnal S. Q. Zhou, Phase selection in Mn-Si alloys by fast solid-state reaction with enhanced skyrmion stability, *Adv. Funct. Mater. 8, 2009723* (2021). The TEM measurements were done by R. Hübner. All other measurements were done by the book author. The book author also analyzed all data and wrote the manuscript.

4.1 Introduction

Magnetic skyrmion are topologically protected spin configurations [166, 167] and have been experimentally discovered in a series of magnetic compounds with noncentrosymmetric crystal structure, such as B20-type transition-metal silicides or germanides [5, 44, 140]. In those materials, the Dzyaloshinskii Moriya (DM) interaction and Heisenberg exchange interaction compete with each other, resulting in the formation of a whirling spin structure. The

57

size of skyrmion or the associated magnetic modulation period is determined by the ratio between the magnitudes of the DM interaction and the ferromagnetic exchange interaction, and typically ranges from 1 to 100 nm [20, 44, 139-141]. To drive skyrmion, a much smaller current is needed in comparison to those used to generate spin transfer torques in ferromagnetic metals, which results in lower power consumption and decreases the Joule heat [23, 143, 168]. Magnetic skyrmion are proposed for applications in new-generation data storage [40, 140].

B20-type MnSi is regarded as the canonical example, in which magnetic skyrmion were firstly observed [5]. Up to now, though bulk B20-MnSi compounds are relatively easy to fabricate by rod-casting furnace or Czochralski method [96, 169], skyrmions are only stable in very narrow temperature and magnetic-field ranges. Thin films of B20-MnSi exhibit a much broader temperature range for the existence of skyrmion [72]. Moreover, the preparation of thin film materials on technology-relevant substrates (such as Si) is a prerequisite for today's information technology. B20-type MnSi films have been prepared by molecular beam epitaxy (MBE) and solid-phase epitaxy (SPE) [101, 102, 105, 170]. However, several scientific issues remain challenging for the MnSi thin-film growth. The most important one is the coexistence of the $MnSi_{1.7}$ impurity phase always with MnSi. Another one is that the film can only be grown to a limited thickness of around 30 nm; for larger film thicknesses the stability of skyrmion is deteriorated. From the equilibrium Mn-Si phase diagram, it is found that the B20-MnSi phase is very sensitive to the ratio of the Mn and Si composition [171]. Diffusion induced ingredient fluctuations will result in the formation of $MnSi_{1.7}$. Moreover, $MnSi_{1.7}$ has a lower crystallization temperature than MnSi [84, 116], and grows naturally as a parasitic phase in the heating or cooling process during the solid-state reaction. $MnSi_{1.7}$ is a weak itinerant magnet and exhibits excellent thermoelectric properties [148, 172]. Recently, it was suggested that $MnSi_{1.7}$ can apply strain to MnSi and increase its Curie temperature [97]. Therefore, separating the MnSi and $MnSi_{1.7}$ phases and/or controlling their mixture ratio is challenging, but is necessary in order to optimize the functionality and to design the topological magnetic properties of MnSi films on demand. This is exactly the question to be tackled in this article.

The relationship between the nucleation activation energy Q, the heating rate Φ and the crystallization temperature T_p can be expressed by the Kissinger equation [173],

$$\ln\left(\frac{\Phi}{T_p^2}\right) = -\frac{Q}{RT_p} + C \tag{17}$$

where R is the gas constant, and C is a constant ($C = \frac{AR}{Q}$, A is pre-exponential factor also known as the frequency factor). For a particular crystalline phase, when increasing the heating rate Φ, the crystallization temperature will be shifted to higher temperatures. The interval of the crystallization temperatures for different phases will be enlarged. So, different phases can be separated and the microstructure can be selected. Moreover, in the situation of ultra-fast heating, the temperature increases so fast that the phase with lower crystallization temperature can not nucleate within such a short time. This has been demonstrated for CuZr-based metallic glasses in Ref [174]. In particular, by increasing the heating rate above 250 Ks^{-1}, the ductile B2 phase has been selectively formed and the formation of the low temperature brittle phases $CuZr_2$ and $Cu_{10}Zr_7$ is suppressed. The fast heating and cooling rates make the system-temperature cross the crystallization temperature of brittle phases in a very short time. There is no nucleation time for the brittle phases due to the transient heating and cooling processes. This approach also works well for amorphous Au-based and Pt-based alloys [175, 176]. Therefore, we anticipate that a fast annealing method, such as flash-lamp annealing (FLA) [124], provides an effective way to separate B20-MnSi and $MnSi_{1.7}$ and to control the ratio between both phases. During FLA, the sample is exposed to xenon flash lamps, such that the total energy budget is low, leading to a fast heating and cooling rate. The details can be found in the supplementary materials.

Here, we report the fabrication of phase-controlled $MnSi_x$ films by the solid-state reaction of metallic Mn layers with Si during millisecond flash-lamp annealing. By controlling the energy density deposited to the sample surface by the flash lamps, single-phase B20-MnSi, $MnSi_{1.7}$ or their controlled mixture can be fabricated. The obtained B20-MnSi thin film has a high Curie temperature of 41 K and exhibits characteristic signatures of magnetic skyrmion. The formation window of skyrmion in our film is significantly enlarged with respect to the magnetic field (2-10 kOe) or temperature (up to 41 K) range compared to previously published results.

4.2 Experiment

Sample preparation: In order to fabricate the MnSi films, a 30 nm thick Mn film was firstly deposited on a Si (111) wafer by DC magnetron sputtering. Afterwards, FLA was employed to anneal the samples at different annealing parameters, rendering the reaction between Mn and Si. The FLA device is made by Rovak GmbH, and illustrated in Figure 5.1.

During annealing, the samples were heated up by 12 Xe lamps of 30 cm long in a continuous N_2 flow. The annealing temperature was controlled by varying the power density. 4.2R, 4.3R and 4.3F represent the samples with annealing parameters of 4.2 kV (Voltage applied to the capacitor of the flash lamps, to produce an intense flash pulse) from the rear side (4.2R), 4.3 kV from the rear side (4.3R) and 4.3 kV from the front side (4.3F), respectively. For all samples, the flash duration was 20 milli-seconds. 4.2 or 4.3 kV is the high voltage to charge the flash lamps, which corresponds to an energy density of 134.6 Jcm^{-2} and 139.5 Jcm^{-2}. Generally, the higher voltage generates the higher temperature on the surface if the absorbance and the materials are the same. The annealing from the front side (4.3F) is supposed to generate a slight lower temperature than 4.3R due to the large reflectivity of the Mn metal film than the Si wafer. With a 20 ms pulse duration, the heating and cooling rates are estimated to be 80000 and 160 Ks^{-1}, respectively.

Therefore, there are three samples included in this chapter. They are

4.2R: low processing temperature, with only $MnSi_{1.7}$;

4.3F: middle processing temperature, with a mixture of $MnSi_{1.7}$ and MnSi;

4.3R: high processing temperature, with only MnSi.

In Chapter 2, we have detailed introduction about the FLA system. Briefly, a reflector, a bank of Xe flash lamps, a wafer holder, two pieces of quartz plate and a preheating module constitute the FLA system. The 12 side-by-side connected Xe lamps can provide a uniform flash across a large area having the size of a 5-inch wafer. To protect the sample from oxidation, continuous N_2 flow is provided during the FLA process. The reflector is designed to obtain a uniform temperature of the annealing chamber. The sample processing chamber consists of two sheets of quartz panels for near ultraviolet and visible light, which separates the air and the protective atmosphere filled inside. By applying a high voltage, the flash lamp is charged and discharged, producing high-density light. By absorbing the light, the temperature of the sample can reach above 1000 K in millisecond time scale. This short time process leads to high heating and cooling rates.

Structure characterization: X-ray diffraction (XRD) and transmission electron microscopy (TEM) were employed to analyse the microstructure of the obtained films. XRD was performed at room temperature on a Brucker D8 Advance diffractometer with a Cu-target source. The measurements were done in Bragg-Brentano-geometry with a graphite secondary monochromator and a scintillator detector. Bright-field and high-resolution TEM imaging were performed on an image-Cs-corrected Titan 80-300 microscope (FEI) operated at an accelerating voltage of 300 kV. High-angle annular dark-field scanning transmission electron microscopy (HAADF-STEM) imaging and spectrum imaging analysis based on energy-dispersive X-ray spectroscopy (EDXS) were done at 200 kV with a Talos F200X microscope equipped with an X-FEG electron source and a Super-X EDXS detector system (FEI).

Magnetic and transport property measurements: The magnetization of the films was measured by a superconducting quantum interference device equipped with a vibrating sample magnetometer (SQUID-VSM) with the field parallel (in-plane) or perpendicular (out-of-plane) to the films. For the zero field cooling (ZFC) measurements, the samples were cooled down to 5 K under a zero field, then different fields were applied and magnetization data was collected during the warming up process. When reaching 100 K, the samples were re-cooled to 5 K at the same field while the data recording was continued. This process was called field cooling (FC). The transport properties of the MnSi film were investigated by a Lake Shore Hall measurement system. Magnetic-field dependent resistance was measured between 5 and 300 K using the van der Pauw geometry. In the resistivity measurement the magnetic field was applied along the sample surface plane (in-plane).

4.3 Results

4.3.1 MnSi and MnSi$_{1.7}$ phase reaction

As shown in Figure 4.1 (a), depending on different annealing temperatures, two reactions are possible. Due to the ultra-fast FLA process, diffusion and nucleation happen at the same time. The diffusion of Si dominates this phase formation process [177]. The reaction at low temperature is the reaction between Mn and Si, leading to MnSi$_{1.7}$ directly. The reaction at high temperature can form B20-MnSi at the beginning. However, since the Si wafer can provide endless Si atoms, MnSi can further react with extra Si, forming the Si-rich phase, namely MnSi$_{1.7}$. In the conventional solid-state reaction process, MnSi$_{1.7}$ is more easily formed as an inevitable secondary phase in MnSi films. According to our anticipation, a fast or transient reaction between Mn and Si by flash lamp annealing can inhibit the formation of MnSi$_{1.7}$ by controlling different annealing temperatures with ultra-fast heating/cooling rates. In the following, we focus on 3 samples: 4.2R, 4.3R and 4.3F, which were annealed using flash lamps from the rear-side (R) or from the front-side (F) with the capacitor charged up to 4.2 kV or 4.3 kV corresponding to the energy density of 134.6 Jcm^{-2} or 139.5 Jcm^{-2}. Therefore, sample 4.2R experienced a lower reaction temperature than sample 4.3R. Sample 4.3F was processed at an intermediate temperature due to the large reflectance of the metallic Mn front side compared to the rough rear side of the Si wafer. In particular, the larger reflectance of the Mn film leads to lower absorbance and therefore lower energy annealing temperature.

Figure 4.1 (b) displays the XRD patterns obtained from the samples prepared at different annealing parameters. The (111) and (222) diffraction peaks of the Si substrate are at 28.4° and 58.9°, respectively. For sample 4.3R (red, bottom trace) annealed at high temperature, the MnSi-(111) and (222) Bragg peaks are present at 34.2° and 72.4°, respectively. MnSi (210) is also observed at 45° with a much weaker intensity. According to the powder PDF card (n. 01-081-0484) [178], the MnSi (210) peak should be the strongest. Taking into account the intensity ratios between the different diffraction peaks, the majority of the MnSi phase in sample 4.3R is (111) textured. In comparison to sample 4.3R, sample 4.2R (blue, top trace) shows only the Bragg peaks of $MnSi_{1.7}$. The lower annealing temperature for sample 4.2R promotes the formation of the $MnSi_{1.7}$ single phase. Sample 4.3F (black, middle trace) was processed at an intermediate temperature and shows a mixture of MnSi and $MnSi_{1.7}$. The MnSi phase is also highly (111)-textured. Thus, by tuning the FLA parameters, we have full control over the Mn-silicide phase formation: single-phase B20-MnSi and $MnSi_{1.7}$ are obtained for the highest and lowest annealing temperatures, respectively, which is in accordance with the Kissinger equation.

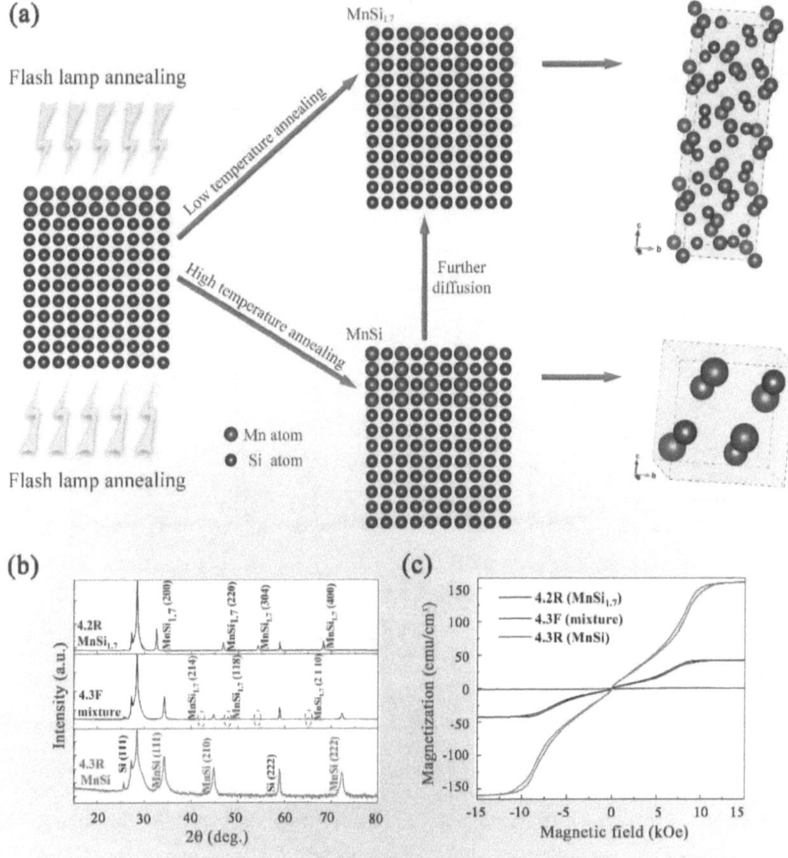

Figure 4.1. *The separation of MnSi and MnSi₁.₇ phases. (a) Schematic representation of atom diffusion and solid state reaction between metal Mn and Si. The blue and purple balls stand for Si and Mn atoms, respectively. MnSi₁.₇ can be formed at low temperature. B20-MnSi is formed at high temperature. MnSi can further react with Si forming MnSi₁.₇ during cooling down. (b) XRD patterns of Mn/Si samples after FLA at different annealing parameters. Single phase MnSi (bottom, red trace) or MnSi₁.₇ (top, blue trace) was formed in sample 4.3R or 4.2R respectively. Sample 4.3F (an enlarged XRD pattern is shown in the supplementary information) is a mixture of MnSi or MnSi₁.₇. (c) In-plane magnetic hysteresis curves recorded at 5 K for the same set of samples. As expected, the magnetization is the highest for the sample containing only B20-type MnSi. The sample containing only MnSi₁.₇ shows negligible*

magnetization. By the transient reaction between Mn and Si via milli-second flash lamp annealing, we can synthesize single phase MnSi or MnSi$_{1.7}$, as well as their mixture at will.

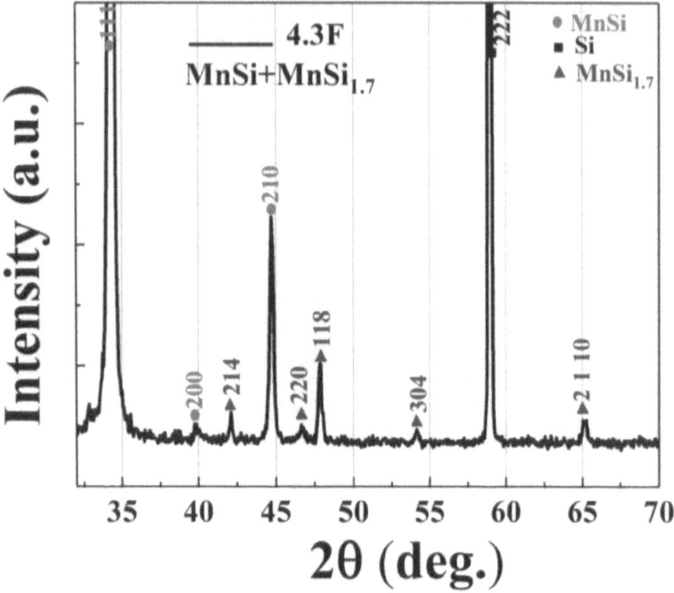

Figure 4.2. Enlarged XRD patterns of sample 4.3F. The XRD pattern shows a mixture of MnSi or MnSi$_{1.7}$.

To better display the diffraction peaks of MnSi$_{1.7}$ from samples 4.3F, the enlarged XRD patterns are shown in Figure 4.2. Sample 4.3F has the mixture of MnSi and MnSi$_{1.7}$. The MnSi phase in this sample is (111) textured. MnSi$_{1.7}$ is polycrystalline without preferable orientation.

The in-plane field-dependent magnetizations of these three samples are shown in Figure 4.1 (c). Sample 4.3R with single B20-MnSi phase displays multi hysteresis, which originates from the transition of different magnetic structures, a typical magnetic feature of B20-MnSi. The sample annealed at 4.3F reveals much reduced magnetization due to the fact that it also contains MnSi$_{1.7}$. As is well known, MnSi$_{1.7}$ is a weak itinerant magnet with a saturation magnetization of 0.012 μB/Mn [82]. As expected, sample 4.2R containing only MnSi$_{1.7}$ shows negligible magnetization in comparison with the other two samples.

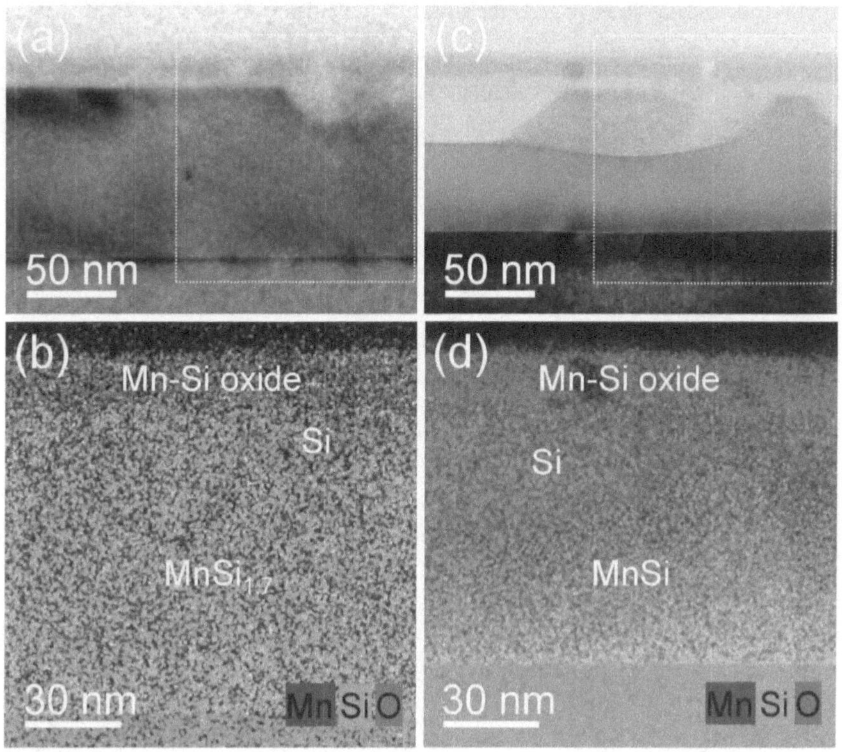

***Figure 4.3.** Cross-sectional bright-field TEM images (a, c) and EDXS-based element distributions (b, d; Mn: blue, Si: green, O: red) for the samples 4.2R (left) and 4.3R (right) confirming the presence of a continuous single-phase MnSi$_{1.7}$ and MnSi film, respectively, each characterized by a sharp interface to the Si substrate. Thus MnSi$_{1.7}$ and MnSi thin film can be synthesized at will by controlling the flash energy density.*

Figure 4.3 (a-b), and (c-d) show cross-sectional bright-field transmission electron microscopy (TEM) images (a, c) and energy-dispersive X-ray spectroscopy (EDXS) based element distributions obtained in scanning TEM mode (b, d) for the samples 4.2R and 4.3R, respectively. According to Figs. 4.3 (a-b), sample 4.2R is characterized by a continuous MnSi$_{1.7}$ film with a thickness of around 90 nm and a sharp interface to the Si substrate. Besides an amorphous Mn-Si oxide capping layer and isolated crystalline Si inclusions between this oxide and the MnSi$_{1.7}$ film, there is no other Mn-Si compound. In contrast, for sample 4.3R, an up to 75-nm-thick continuous single-phase B20-type MnSi film is obtained (Figure 4.3 (c-d)). Moreover, extra Si from the substrate can diffuse through the MnSi layer, and an amorphous

Mn-Si oxide capping layer protects the sample from further oxidation. The TEM analyses, which are described in more detail in the Figure 4.4 and 4.5, well corroborate the XRD and magnetization measurements. By controlling the reaction temperature during millisecond flash lamp annealing, we can obtain films of pure tetragonal $MnSi_{1.7}$, of pure cubic B20-type MnSi or of their mixture.

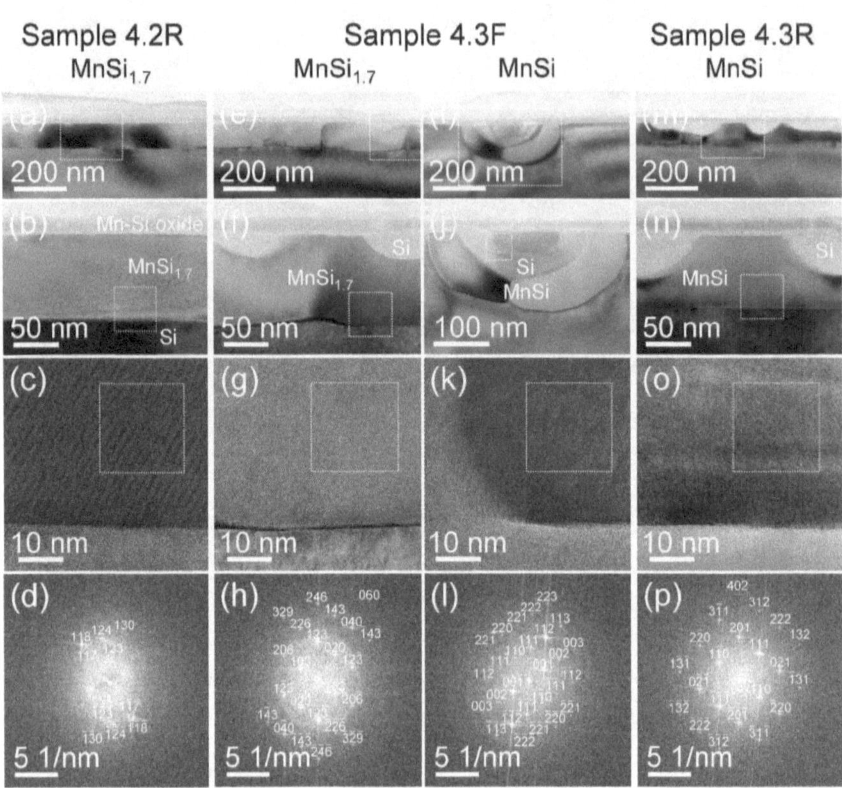

Figure 4.4. Structure and morphology characterisation of regrown films at different annealing parameters. . Cross-sectional bright-field TEM images at low (1st row: a, e, i, m) and medium magnification (2nd row: b, f, j,, n) as well as high-resolution TEM images (3rd row: c, g, k, o) and corresponding FFTs (4th row: d, h, l, p) for the samples 4.2R (1st column: a-d), 4.3F (2nd & 3rd column: e-l), and 4.3R (4th column: m-p). For each sample position, the regions where the TEM image at increased magnification, the HRTEM image, and the associated FFT, respectively, were taken are marked in the correspondingly preceding panels with dashed squares. All micrographs were recorded in Si [$\bar{1}$ 1 0] zone axis geometry, except for panels (o-

l) which were taken in MnSi [1 1 $\bar{2}$] zone axis geometry and hence about 6 degree off [$\bar{1}$ 1 0] Si. Each FFT was indexed using the JEMS software package: (d) close to MnSi$_{1.7}$ [12 $\bar{4}$ $\bar{1}$] zone axis pattern, (h) MnSi$_{1.7}$ [3 0 $\bar{1}$] zone axis pattern, (l) MnSi [$\bar{1}$ 1 0] zone axis pattern, (o) MnSi [1 1 $\bar{2}$] zone axis pattern.

Figure 4.4 (a-d), (e-l), and (m-p) show cross-sectional TEM images of the samples 4.2R, 4.3F and 4.3R, respectively. While the panels (a, e, i, m) and (b, f, j, n) display overview and more detailed bright-field TEM images, panels (c, g, k, o) and (d, h, l, p) depict high-resolution TEM micrographs and corresponding fast Fourier transforms (FFT), respectively. As shown in Figure 4.4 (a-c), sample 4.2R has a continuous MnSi$_{1.7}$ film with a thickness of around 90 nm and a sharp interface to the single-crystalline Si substrate. The FFT in Figure 4.4 (d) can be described best with a pattern close to the [12 $\bar{4}$ $\bar{1}$] zone axis (in particular, normal to the growth plane [1 2 4] MnSi$_{1.7}$ is parallel to [1 1 1] Si and in-plane [12 $\bar{4}$ $\bar{1}$] MnSi$_{1.7}$ is almost parallel to [$\bar{1}$ 1 0] Si). Besides an amorphous Mn-Si oxide cap layer and crystalline Si inclusions between this oxide and the MnSi$_{1.7}$ film, there is no other Mn-Si compound. Figure 4.4 (e-l) show TEM images of sample 4.3F, confirming the presence of both phases MnSi$_{1.7}$ (Figure 4.4 (e-h)) and B20-type MnSi (Figs. 4.4 (i-l)). For this sample, the MnSi$_{1.7}$ phase grows as an approximately 100-nm-thick continuous film with grains of various orientations (e.g. normal to the growth plane [1 2 3] MnSi$_{1.7}$ is parallel to [1 1 1] Si and in-plane [3 0 $\bar{1}$] MnSi$_{1.7}$ is parallel to [$\bar{1}$ 1 0] Si (Figure 4.4 (h)), while the MnSi phase is only found for isolated islands (normal to the growth plane [1 1 1] MnSi is parallel to [1 1 1] Si and in-plane [$\bar{1}$ 1 0] MnSi is parallel to [$\bar{1}$ 1 0] Si (Figure 4.4 (l)). With further increasing reaction temperature for sample 4.3R, an up to 75-nm-thick continuous single-phase B20-type MnSi film is obtained (Figure 4.4 (m-p)). For the analysed TEM region, normal to the growth plane [2 0 1] MnSi is parallel to [1 1 1] Si and in-plane [1 1 $\bar{2}$] MnSi is about 6 degree off [$\bar{1}$ 1 0] Si (Figure 4.4 (p)). The TEM analyses well corroborate the XRD and magnetization measurements.

To obtain two-dimensional element distributions and confirm the above-described structure characterizations, high-angle annular dark-field scanning TEM (HAADF-STEM) imaging combined with spectrum imaging analysis based on energy-dispersive X-ray spectroscopy (EDXS) were performed. While panels (a, c, e, g) of Figure 4.5 give the HAADF-STEM micrographs, panels (b, d, f, h) show the corresponding superimposed element maps (Mn: blue, Si: green, O: red). The positions analysed for sample 4.3F (Figure 4.6 (c-f) are exactly the same as in Figure 4.5, while the chosen regions for sample 4.2R (Figure 4.5 (a-b)) and 4.3R (Figure 4.5 (g-h)) are slightly different. The compositional maps are in accordance with the TEM-based analysis results of Figure 4.5. In particular, quantitative chemical analysis of the Mn-Si films gives values very close to the $MnSi_{1.7}$ and B20-type MnSi phases.

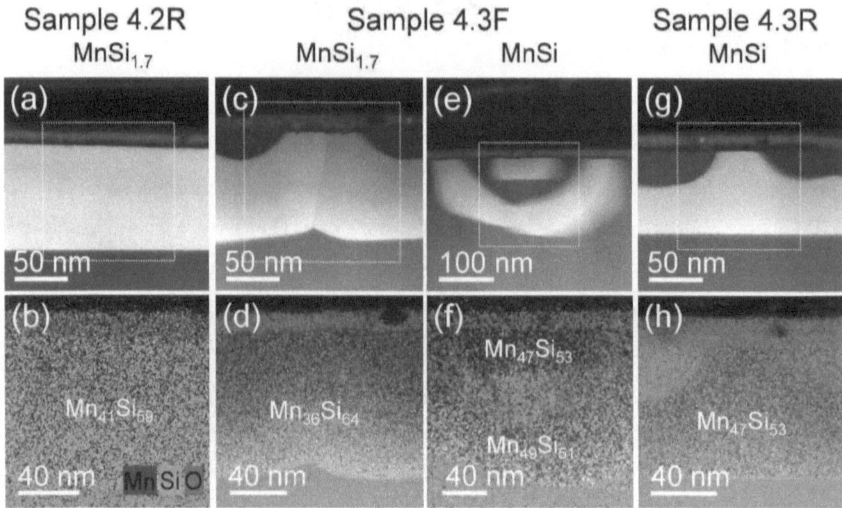

Figure 4.5. Chemical composition characterisation of regrown films at different annealing parameters. Cross-sectional HAADF-STEM images (1ˢᵗ row: a, c, e, g) and superimposed element maps (2ⁿᵈ row: b, d, f, h) for the samples 4.2R (1ˢᵗ column: a-b), 4.3F (2ⁿᵈ & 3ʳᵈ column: c-f), and 4.3R (4ᵗʰ column: g-h). For each sample position, the region where the EDXS-based element maps (Mn: blue, Si: green, O: red) were obtained are marked in the corresponding HAADF-STEM image with a dashed square. The results of quantitative chemical analysis are given for the various Mn-Si regions.

4.3.2 Magnetic Skyrmion

Lorentz transmission electron microscopy (LTEM), magnetic force microscopy (MFM) and neutron scattering can give direct confirmation of magnetic skyrmions in real or reciprocal space [179]. For MnSi thin films, both LTEM and magnetization measurements were performed on the same sample and provided consistent results. Therefore, magnetization and magneto-transport measurements are also accepted as methods to characterize magnetic skyrmions [7, 72, 101]. In this section, we present a detailed magnetic and electrical investigation of sample 4.3R, which contains only the B20-MnSi phase. The MnSi film prepared by our method exhibits non-trivial magnetic properties as known from samples prepared by other methods [7, 36, 72, 101].

Figure 4.6 (a) shows the temperature-dependent saturation magnetization for the B20-MnSi film. The derivative dM/dT of the MT curve has a minimum at 41 K, indicating the Curie temperature, which is higher than for bulk MnSi (29.5 K [5]) due to the tensile strain in the MnSi film [108] (for details see Raman analysis in Figure 4.7 [115, 157, 158]). The ZFC/FC magnetization under different magnetic fields are shown in Figure 4.6 (c). Near 41 K, a cusp feature is observed. It is the transition from the ground helimagnetic to paramagnetic state above 41 K. When increasing the magnetic field, the cusp becomes more pronounced. The shift and disappearance of this cusp are shown in Figure 4.8, which prove the transformations of different magnetic structures again.

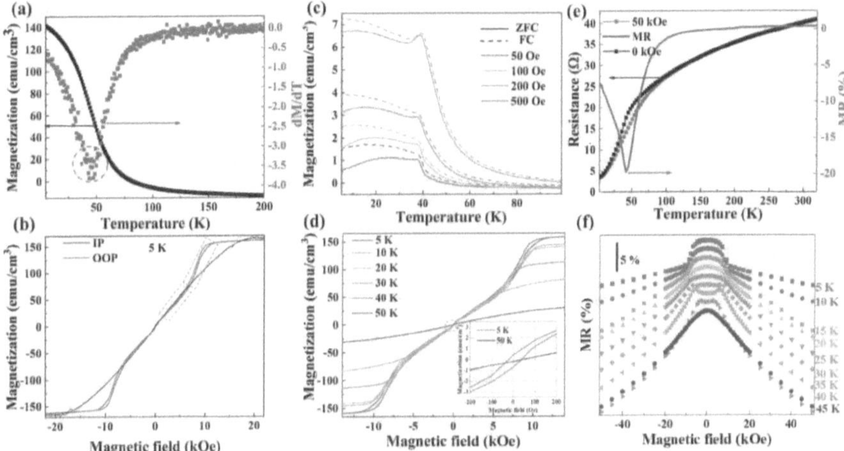

Figure 4.6. *(a) In-plane saturation magnetization and the calculated dM/dT. The valley of dM/dT at around 41 K indicates the Curie temperature or the transition from ferromagnetic to paramagnetic. (b) In-plane and out-of-plane MH curves recorded at 5 K. The easy axis and multi-hysteresis are stabilized in-plane. The area of green dotted rectangles show multi-hysteresis. (c) In-plane field cooling (FC) and zero field cooling (ZFC) magnetization measured under different magnetic fields. There is also a kink around 41 K, which is the transition from helical to paramagnetic. (d) In-plane MH curves measured at different temperatures. The inset shows the data around the origin. (e) Temperature-dependent resistance under 0 (black square) and 50 kOe (red circle) fields. The red solid line is the calculated MR at 50 kOe (f) Magnetic field-dependent magnetoresistance at various temperatures. The anomalous behaviour disappears at 45 K (above the Curie temperature).*

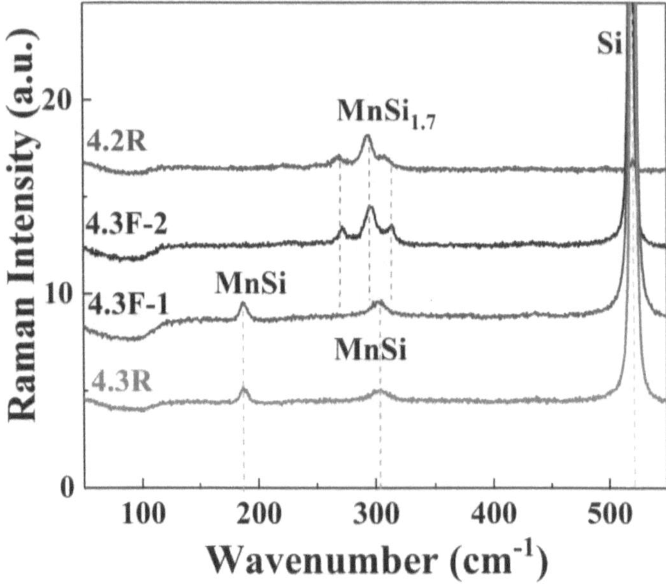

Figure 4.7. Room-temperature Raman spectra for different samples. For each sample, the Raman spectra were measured at 6 positions. For sample 4.3R, different areas show the same Raman signal, pointing to a homogeneous MnSi phase distribution. Sample 4.3F shows two sets of spectra, indicating a mixtured MnSi and MnSi₁.₇ phases. Sample 4.2R also shows the same Raman spectrum at different positions, indicating the homogeneous MnSi₁.₇ distribution.

Figure 4.7 shows the room-temperature Raman spectra of three different samples, 4.2R, 4.3F and 4.3R. The experiments were done using a Horiba micro-Raman system with the excitation wavelength of 532 nm and the signal was recorded with a liquid-nitrogen-cooled silicon CCD camera. The peak at around 520.5 cm^{-1} corresponds to the transverse/longitudinal optical (TO/LO) phonon mode from the Si substrate. The difference between the Raman spectra of three samples provides valuable information about the different crystalline phases. Each sample was checked at 6 different positions. Sample 4.3R shows two well-separated peaks at about 188 and 304 cm^{-1}, which are very probably corresponding to the T_2 Raman-active phonon modes in MnSi. In fully relaxed bulk MnSi, these two phonon modes are located at about 194 and 316 cm^{-1}. The Raman spectroscopy performed in the backscattering geometry provides information about the in-plane strain and lattice vibration. The change of the phonon mode position with respect to the relaxed material indicates tensile strain for a red-shifted spectrum and compressive strain for a blue-shifted spectrum. In our case, the shift of the

70

phonon mode positions towards lower wavenumber indicates the existence of the in-plane biaxial tensile strain. The Raman spectrum obtained from sample 4.2R exhibits three separated peaks at around 300 ± 5 cm^{-1}. These peaks confirm the formation of the MnSi$_{1.7}$ phase. These three Raman peaks should account for the slightly different neighbourhoods of the Mn–Si bonds. At different positions of sample 4.3F, two different spectra were recorded. One is similar to sample 4.3R, pointing the existence of the MnSi phase. The other is similar to sample 4.2R, indicating the presence of the MnSi$_{1.7}$ phase.

Magnetic hysteresis loops at 5 K of the up to 75 nm thick MnSi film with in-plane and out-of-plane magnetic fields reveal the in-plane easy axis favoured by the shape anisotropy [115], as shown in Figure 4.6 (b). Moreover, both hysteresis loops show almost zero remanence, indicating the formation of a multi-domain state at zero field. The saturation field is about 13 and 19 kOe for the in-plane and out-of-plane direction, respectively. In addition, the multi-hysteresis feature is observed only in-plane, indicating Bloch-type Skyrmion [180]. Figure 4.6 (d) shows in-plane MH curves at various temperatures, where the MH curves have multi hysteresis below the Curie temperature. This suggests the transformation between different spin structures. For better visualization, enlarged MH curves from -200 Oe to 200 Oe are shown in the insert of Figure 4.6 (d). Magnetic hysteresis disappears above the Curie temperature, where the material becomes paramagnetic.

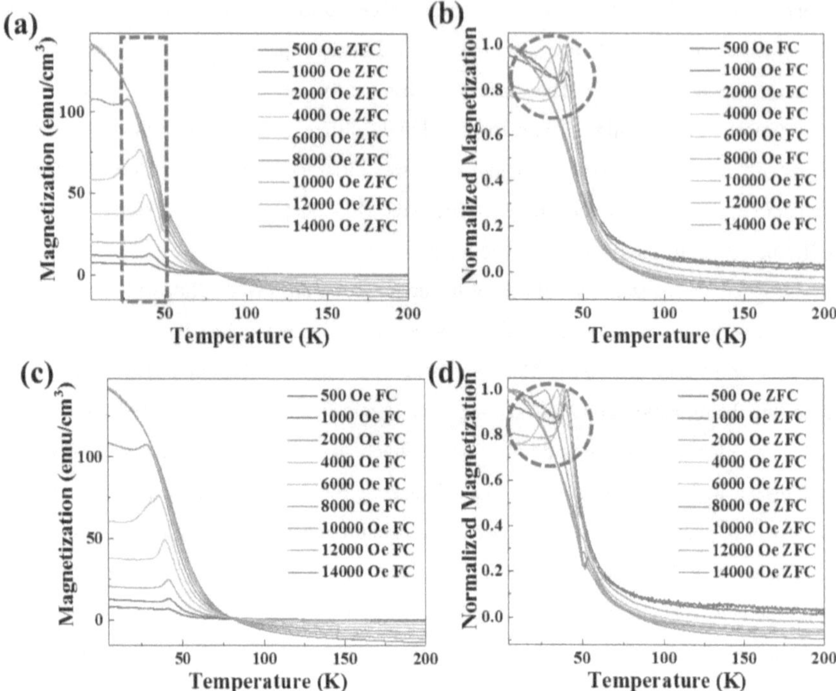

Figure 4.8. Magnetic properties of sample 4.3R. (a) Zero field cooling (ZFC) magnetization curves (b) Normalized zero field cooling (ZFC) magnetization curves measured under different magnetic fields. (c) Field cooling (FC) magnetization curves and (d) normalized Field cooling (FC) magnetization curves measured under different magnetic fields. Below 2000 Oe, the cusps (show by dashed rectangle and circle) in the magnetization curves keep the same temperature of 41 K. It is because the materials stay in full helical state. With increasing the magnetic field, this cusp shifts to lower temperature because the helical structure dissolves gradually. Over 8000 Oe, the cusp disappears since there is no helical structure anymore.

The ZFC and FC curves of the sample 4.3 R are shown in Figure 4.8. The ground state of B20 MnSi is helical. There is a clear cusp in ZFC and FC curves, indicating the transition of the helical phase to another phase. It is found that the cusp does not shift to lower temperature when the magnetic field is below 2000 Oe, because the helical phase exists in the whole temperature range below the transition temperature at these fields. If the applied field is higher than 2000 Oe, the cusp moves to a lower temperature and it disappears when the magnetic field is higher than 8000 Oe due to the fact that at these magnetic fields, the helical phase is not stable. This is consistent with the magnetic phase diagram as shown in the main manuscript. The helical phase does not disappear below $H_{\alpha1}$ and it does not shift below H_β. To better visualize this trend, the normalized ZFC and FC curves are calculated in Figure 4.8 (b) and (d).

The electrical resistances of the MnSi film under a magnetic field of 0 and 50 kOe are shown as a function of temperature in Figure 4.6 (e). The resistance increases with increasing temperature, which means that the MnSi film behaves like a metal [162]. The calculated magnetoresistance (MR), $MR = \frac{R_H - R_0}{R_0} \times 100\%$ (R_H: resistance at 50 kOe, R_0: resistance at zero field), is shown as a red solid line in Figure 4.6 (e). The negative magnetoresistance in MnSi can be understood as follows: the magnetic field increases the effective field acting on the localized spins and suppresses the fluctuation of spins in space and time, which leads to a decrease of the resistivity. The peak of (negative) MR at around 41 K implies that the largest resistance change occurs near the Curie temperature.

The magnetoresistance of the MnSi film at various temperatures is shown in Figure 4.6 (f). Below the Curie temperature, it shows an anomalous phenomenon at low magnetic field. More MR data around the Curie temperature is presented in Figure 4.9. Above the Curie temperature, the MR becomes normal without any specific features, which is consistent with other MnSi films.

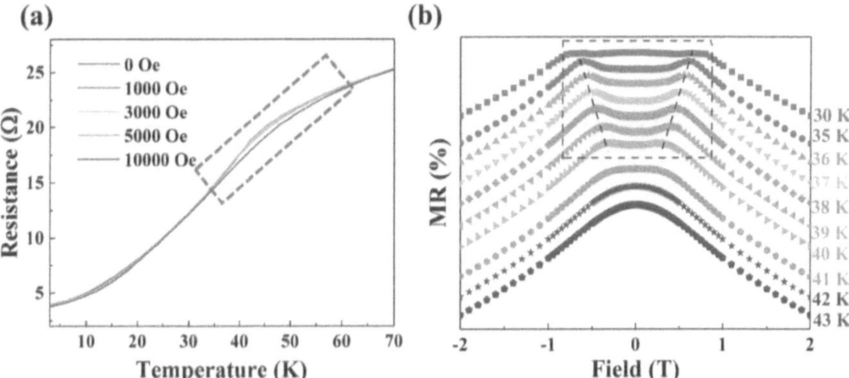

Figure 4.9. (a) Temperature-dependent resistance at different magnetic fields and (b) magnetic-field-dependent magnetoresistance at various temperatures. The variation of resistance (shown by dashed rectangle) around the Curie temperature is becoming obvious with increasing the magnetic field. The anomalous behaviour of magnetoresistance disappears at 41 K (Curie temperature).

Figure 4.9 (a) shows the temperature-dependent in-plane resistance at different magnetic fields. There is an obvious variation (indicated by the dashed rectangle) around the Curie temperature. With increasing magnetic field, the resistance decrease strongly. The magnetic-field-dependent magneto-resistance (MR) of sample 4.3R is shown in Figure 4.9 (b). Below the Curie temperature, it shows an anomalous behaviour below the saturated magnetic field (shown by the dashed rectangle). Above the Curie temperature, the MR is negative and without any specific features. The anomalous behaviour becomes smaller with increasing temperature (shown by the dashed lines), indicating the critical fields shifting to lower magnetic field. This is well collaborated with the magnetic phase diagram in the main manuscript.

In Figure 4.10 (a) and (c), the derivatives of the static magnetization dM/dH show four critical transition fields, termed as H_β, $H_{\alpha 1}$, $H_{\alpha 2}$ and H_{sat}. The dashed lines show their shift depending on temperature. H_{sat} is the critical field, above which the material changes into field-polarized ferromagnetism. Below H_β, the system is at its ground helicoid state. At 40 K, there is no clear peak resolvable, but only a plateau, meaning the smearing out of phase boundary near the Curie temperature. With decreasing temperature, we observe two apparent peaks at $H_{\alpha 1}$ and $H_{\alpha 2}$, indicating the appearance of another phase. The first-order transitions in and out of this phase bounded by $H_{\alpha 1}$ and $H_{\alpha 2}$ indicate a difference in topology between this state and the neighbouring ferromagnetic (above $H_{\alpha 2}$) and helicoid (below $H_{\alpha 1}$) states. The phase between $H_{\alpha 1}$ and $H_{\alpha 2}$ is the so-called skyrmion phase. Above $H_{\alpha 2}$ the skyrmion phase nearly vanishes and the system evolves via a first-order phase transition into a conical state. With increasing field from H_β to $H_{\alpha 1}$, the system is in a metastable helicoid state. Compared with field-increase and decrease processes, these transition fields show small difference below 15 K due to the hysteresis. However, above 15 K these transition fields do not change significantly with respect to field increase or decrease due to the coinciding of MH curves.

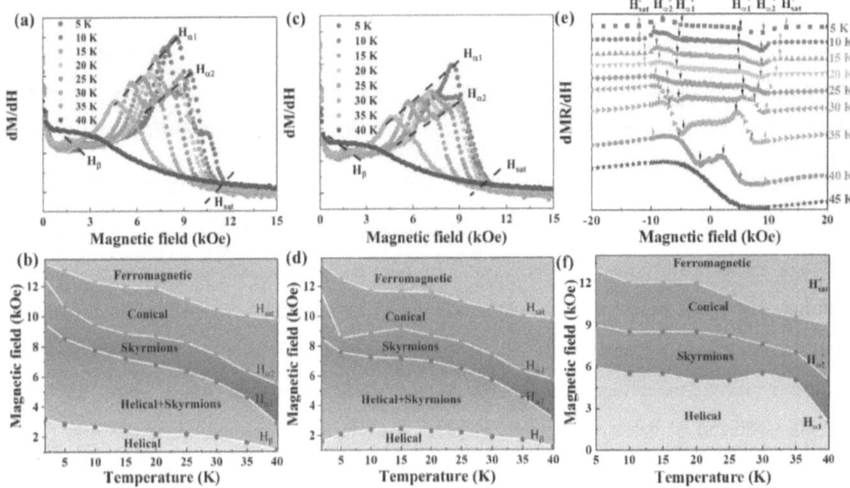

Figure 4.10. *Static susceptibility in increasing (a) and decreasing (c) field sweeps for the MnSi film at various temperatures. Black dashed lines in (a) and (c) show the shift of transition fields at each temperature. Four critical fields are identified and labelled as H_β, $H_{\alpha1}$, $H_{\alpha2}$, and H_{sat}. $H_{\alpha1}$ and $H_{\alpha2}$ bound the region of stable elliptic skyrmion. The dashed lines define the boundary for metastable helicoids H_β and ferromagnetic H_{sat}. Magnetic phase diagram in increasing (b) and decreasing (d) field. With increasing temperature, these magnetic phases (Helical, Helical + Skyrmion, Skyrmion, Conical and Ferromagnetic) can be formed at lower fields. (e) Calculated dMR/dH from data in Fig. 4.6 (f). The black, red and green arrows stand the position of $H_{\alpha1}$, $H_{\alpha2}$ and H_{sat}, respectively. (f) Magnetic phase diagram of the MnSi film with respect to temperatures and magnetic fields.*

The temperature-dependent transition fields are plotted in Figure 4.10 (b) and (d). With increasing the magnetic field, the magnetic structure transforms into the Helical, Helical + Skyrmion, Skyrmion, Conical and Ferromagnetic phase. Compared with other B20-MnSi, Skyrmion in our MnSi film can be stabilized over the whole temperature range below the Curie temperature and a wider magnetic field range of around 10 kOe. It is known that the temperature and magnetic field intervals of skyrmion become narrower with increasing the thickness of MnSi film. However, our MnSi film has a thickness of up to 75 nm and is thicker than other reported MnSi films. This enlarged skyrmion stability in wider temperature and magnetic field ranges offers more flexibility in spintronic applications [23, 26, 168]. It is speculated that the reaction by flash lamp annealing stabilizes the strain above the critical thickness, therefore increasing both the Curie temperature and the stability of skyrmion.

Indeed, by analysing the magnetoresistance one also can investigate the field-driven evolution of the spin textures in MnSi [115, 117]. As shown in Figure 4.10 (e) we examine more closely the derivative dMR/dH for the same data displayed in Figure 4.6 (f). We defined the transition fields $H_{\alpha 1}$, $H_{\alpha 2}$ and H_{sat}, as marked by the arrows. Values for positive and negative magnetic fields are denoted by the superscripts "+" and "−", respectively. However, the transition from Helical to Helical + Skyrmion is not resolved probably due to the negligible MR variation, being consistent with previous reports. With increasing magnetic field, the helimagnetic phase transforms into skyrmion via a first order phase transition, manifested as peaks at $H_{\alpha 1}$ due to the completely different topological properties between helimagnetic and Skyrmion. Above the critical field $H_{\alpha 2}$, the system transforms into the conical phase. Skyrmion can be stabilized at the range between $H_{\alpha 1}$ and $H_{\alpha 2}$. H_{sat} is the critical magnetic field, above which MnSi changes into ferromagnetic state. Figure 4.10 (f) shows the magnetic phase diagram of MnSi. Compared with Figure 4.10 (b) and (d), this magnetic phase diagram obtained from MR shows qualitatively the same temperature and magnetic field range for skyrmion.

4.3.3 Discussion

There are 4 stable Mn-Si compounds. With increasing Si concentration, Mn_3Si, Mn_5Si_3, B20-MnSi and higher manganese silicides can form at thermal equilibrium. Characterized by tetragonal crystal structures with different c-axis lengths, higher manganese silicides have many chemical formulas such as: Mn_4Si_7, $Mn_{11}Si_{19}$, $Mn_{15}Si_{26}$ and $Mn_{27}Si_{47}$ and are generally written as $MnSi_{1.7}$ [181]. The single B20-MnSi phase can only form at 50 at.% Si and $MnSi_{1.7}$ exists from 55 to 95 at.% Si. In the Mn-Si thin film grown on Si substrates, the richness of Si makes the Mn_3Si and Mn_5Si_3 phase hard to form due to the limited amount of Mn. However, $MnSi_{1.7}$ is much easier to form in these thin films. Moreover, the crystallization temperature for B20-MnSi is 1276 °C, which is higher than 1150 °C for $MnSi_{1.7}$. This means that $MnSi_{1.7}$ can nucleate prior to B20-MnSi in thermal equilibrium condition. Therefore, $MnSi_{1.7}$ often coexists with B20-MnSi in thin films grown by solid-phase epitaxy. The equilibrium phase diagram of the Mn-Si system is shown in Figure 4.11 (a).

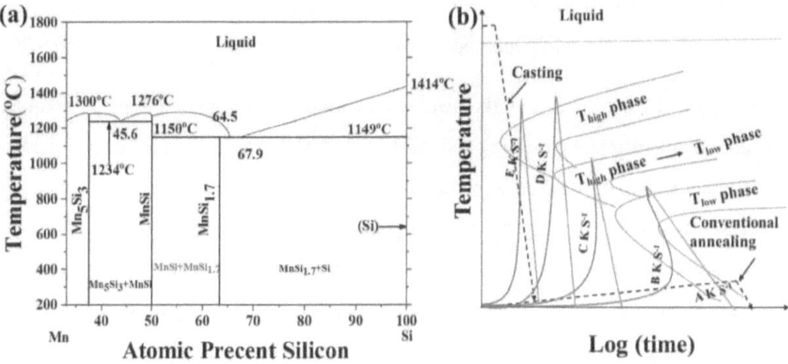

Figure 4.11. (a) The equilibrium Mn-Si phase diagram from Ref 2: Single MnSi phase can only exist at 50 at.% Si; however, MnSi$_{1.7}$ appears in a much broader window. (b) The Continuous Heating Transformations (CHT) diagram from Ref 3. The red curves stand for annealing processes at different heating rates and the green curves stand for cooling processes. The A, B, C, D and E stand for the increasing heating rates. T$_{low}$ phase and T$_{high}$ phase stand for the phase formed at low temperature and high temperature, respectively. When the heating rate is above a critical heating rate D and the temperature is above the formation temperature of the low-temperature phasethe latter can be suppressed. If the cooling process is also very fast, like the annealing process, the low-temperature phase can be avoided.

Figure 4.11 (a) shows the enlarged phase diagram of Mn-Si compounds. Mn$_5$Si$_3$, B20 MnSi and higher manganese silicides can form under thermal equilibrium with increasing the amount of Si. As pure phase, the line compound B20-MnSi can only form at 50 at% Si, and MnSi$_{1.7}$ always coexists with B20-MnSi due to composition fluctuation. Mn$_5$Si$_3$ is hard to form in our thin film due to the richness of Si from the substrate. The crystallization temperature for MnSi$_{1.7}$ is 1150 °C, i.e. lower than that of MnSi (1276 °C), indicating that MnSi$_{1.7}$ can nucleate prior to B20-MnSi during heating in thermal equilibrium condition. Therefore, MnSi$_{1.7}$ often coexists with B20-MnSi in thin films grown by solid phase epitaxy. The phase selection can be realized by controlling the heating and cooling rates. With increasing the heating rate above D Ks^{-1}, the high-temperature phase can be selectively formed, and the formation of low-temperature phases will be suppressed, as shown in Figure 4.11 (b). If the cooling rate is also high enough, the temperature cross the crystallization temperature for low-temperature phase. Then, the low-temperature phase can be avoided. As shown in Figure 4.11 (b), with the heating rate at D Ks⁻¹ and the high cooling rate, only the high-temperature phase can be formed. Furthermore, this fast annealing method could be applied in other B20 material systems.

During flash lamp annealing, the heating and cooling rates are much higher compared to conventional rapid thermal annealing and can be adjusted by controlling the energy budget, i.e. the flash energy density delivered to the sample surface. However, we cannot measure the temperature in-situ. Presumably, by increasing the heating rate above a threshold (see Figure 4.11 (b)), the high-temperature phase can be selectively formed and the formation of low-temperature phases will be suppressed, as schematically shown in Figure 4.11 (b). Therefore, this fast annealing method leads to a transient reaction between Mn and Si, serving as an effective way to separate B20-MnSi and $MnSi_{1.7}$ and to control the ratio between MnSi and $MnSi_{1.7}$ in the presence of both phases. Note that for this purpose the fast reaction and the high heating/cooling rate are important. This approach is not only limited to flash lamps. Millisecond lasers are expected to lead to similar effects and have been widely used in semiconductor industry [182].

Both magnetron sputtering and flash lamp annealing are ready to be scaled up to whole Si wafers. Therefore, this approach can be integrated with the existing and well-developed Si microelectronic technology. As shown in Figure 4.6, our $MnSi_{1.7}$ film has a reasonable flat interface and surface. The surface oxide can be selectively etched by acids (e.g. HF). By applying lithography to pattern the Mn film and therefore the $MnSi_{1.7}$ film, one can integrate a stand-along, thermoelectric power source to Si-based devices on a single chip [183]. The top surface of B20-MnSi layer is still not smooth enough. This might be due to the slight oxidation of the Mn metal film right before annealing. However, the surface can be processed by by well-develop chemical mechanical planarization [184]. Nevertheless, the much higher Curie temperature and the broader magnetic-field and temperature window for stabilizing Skyrmion in our MnSi film are profound advantages. They might be exclusively related with the ultrafast thermal processing.

4.4 Conclusion

In summary, by controlling the reaction parameters using strongly non-equilibrium flash lamp annealing, we have full control over the phase formation of Mn-silicides in thin films from single-phase B20-MnSi or $MnSi_{1.7}$ to mixed phases. The obtained films are highly textured and reveal sharp interfaces to the Si substrate. The obtained B20-MnSi films exhibits a high Curie temperature at 41 K. The Skyrmion phase can be stabilized in broad temperature and magnetic field ranges. We propose flash-lamp-annealing-induced transient reaction as a general approach for phase separation in transition-metal silicides and germinides and for growth B20-type films with enhanced topological stability.

5. On the Curie temperature of MnSi films

B20-type MnSi is the prototype magnetic skyrmion material. Thin films of MnSi have a higher Curie temperature than its bulk counterpart. It is not yet clear by what and how the Curie temperature of MnSi thin films is affected. In this chapter, we grow MnSi films on Si(100) and Si(111) substrates with a broad variation in their structures. By controlling the Mn thickness and annealing parameters, the pure MnSi phase of polycrystalline and textured nature as well as the mixed phase of MnSi and $MnSi_{1.7}$ are obtained. Surprisingly, all these MnSi films show an increased Curie temperature of up to around 43 K. However, the Curie temperature is independent of the structural parameters within our accessibility including the film thickness above a threshold, strain, cell volume and the mixture with $MnSi_{1.7}$. Our work is drawing to revisit the origin of the Curie temperature of MnSi films.

Presented results are under preparation for publication: Z. C. Li, Y. Yuan, V, Begeza, L. Rebohle, M. Helm, K. Nielsch, S, Prucnal S. Q. Zhou, on the Curie temperature of MnSi films. All measurements were done by the book author. The book author also analyzed all data and wrote the manuscript.

5.1 Introduction

Bulk manganese monosilicide (MnSi) is a weak itinerant helical magnet with B20 crystal structure [185]. At ambient pressure, bulk MnSi shows magnetic order with a Curie temperature (T_C) of ~ 29.5 K. It has been under investigation for a few decades regarding its intriguing physical properties, such as magnetic quantum phase transition [186, 187] and the formation of a non-Fermi liquid phase [3, 188]. With respect to practical applications, the most attractive property of MnSi is the formation of a magnetic skyrmion lattice, which is a topologically stable spin configuration and promising for spintronic application. A magnetic skyrmion lattice was experimentally observed in bulk MnSi by Mühlbauer $et\ al.$ by using small angle neutron scattering [5]. This work has greatly motivated the development of MnSi thin films on Si substrates. Generally, MnSi thin films have been prepared by molecular beam epitaxy (MBE) [58, 100, 105, 108, 144], solid-state phase epitaxy [72, 106, 112] and magnetron sputtering [145]. Interestingly, independent of the preparation methods, epitaxial-like MnSi films grown on Si(111) show much enhanced T_C to 35-45 K [72, 106, 108, 145]. On Si(111) substrates, MnSi(111) is rotated by 30° with the orientation relationship of Si(111)‖MnSi(111) and Si[11$\bar{2}$] ‖ MnSi[1$\bar{1}$0], leading to a lattice mismatch of around -3.0% ([a_{MnSi} cos(30°) − a_{Si}]/a_{Si} = -3.0%). This induces an in-plane lattice expansion in the MnSi films. It has been shown experimentally that for bulk MnSi the hydrostatic pressure decreases its T_C [187] while the negative chemical pressure can increase its T_C [161, 189]. Therefore, the increased Tc in MnSi films was presumably attributed to the tensile strain from the mismatch with Si substrate [106, 144]. However, the detailed analysis does not support this assumption, since thinner films with a larger cell volume show lower T_C than thicker films [106].

Karhu *et al.* have systematically checked the change of T_C on MnSi films with different thicknesses and strain [106]. Indeed, it was found out that the thinner films show lower T_C and all thicker (>10 nm) films exhibit a similar T_C at around 43 K. A proportional correlation is observed between the Tc and the ratio between the out-of-plane and the in-plane strain. Li *et al.* also found that the T_C of 50 nm thick MnSi film is almost identical to that of the 10 nm film [72]. López *et al.* found that a 30 nm-thick MnSi film does not develop any long-range magnetic order, while the 150 nm MnSi film has a T_C at 34 K [145]. In general, it is known that with increasing thickness of the thin films, the strain originating from the interface should relax [190-192]. In thicker MnSi films, the T_C is expected to be lower than in thinner films. To understand the relationship between strain, atomic bonds and T_C in MnSi films, Figueroa *et al.* have investigated thick MnSi films by polarization-dependent extended X-ray absorption fine structure and found that the Mn positions are unchanged. They concluded that for thick MnSi films the unit cell volume should be essentially the same as for bulk MnSi. They attributed the enhanced T_C to the interface, whose particular, unidentified characteristics strongly affect the magnetic properties of the entire MnSi film, even far from the interface. However, the very thin (below 5 nm) MnSi film shows lower T_C [106] or the absence of magnetic order [145]. Just recently, Sukhanov *et al.* reported an improved T_C of bulk MnSi lamellae with μm dimensions embedded in $MnSi_x$ (x~1.7) matrix [97]. The lattice mismatch between MnSi lamellae and the $MnSi_{1.7}$ matrix produces a tensile strain in MnSi. To understand the increased T_C, it has been assumed that the interface influences the μm thick lamellae. Therefore, the origin of the increased T_C in MnSi films is still unknown.

Here, we report a systematic investigation on the Curie temperature of MnSi thin films with a large variation in their structural properties. These thin films were prepared by the solid-state reaction of metallic Mn layers with Si during ms-range flash lamp annealing. By controlling the Mn thicknesses and annealing parameters (energy density deposited to the sample surface by flash lamps), pure phase B20-MnSi and its mixture with $MnSi_{1.7}$ are prepared both on Si(100) and Si(111) substrates. All obtained thin films have a high Curie temperature around 43 K and the characteristic signature of magnetic skyrmions. We attempt to find a correlation between the Curie temperature and the structural properties, and therefore shed light on the understanding of the increased Curie temperature in thin films.

5.2 Experiment

In order to fabricate MnSi films, 7-30 nm thick Mn films were firstly deposited on Si(100) and (111) wafers by DC magnetron sputtering. Afterwards, flash lamp annealing (FLA) was employed to realize a fast solid-state reaction between Mn and Si at different annealing parameters. The largest thickness of the MnSi is about 60 nm (see Table 1). During the FLA process, these samples were heated up by 12 Xe-lamps in a continuous N_2 flow. Samples were annealed either from the front side or from the backside with a peak temperature above 1300 K. With a 20 ms pulse duration, the heating and cooling rates are estimated to be around 80000 and 160 Ks^{-1}, respectively. Such high heating /cooling rate will allow the control over the parasitic growth of $MnSi_{1.7}$ in B20-type MnSi. By changing the flash lamp energy (and therefore the peak temperature), we can selectively prepare the pure phases of MnSi and $MnSi_{1.7}$ or their mixture. The details about the preparation have been reported in Ref. [142]. All samples used in this manuscript are summarized in Table 5.1.

X-ray diffraction (XRD) was employed to analyse the microstructure of the obtained films. XRD was performed at room temperature on a Bruker D8 Advance diffractometer with a Cu-target source. The measurements were done in Bragg-Brentano-geometry with a graphite secondary monochromator and a scintillator.

The magnetic properties of the films were measured by a superconducting quantum interference device equipped with a vibrating sample magnetometer (SQUID-VSM) with the field parallel (in-plane) to the films. For measuring the temperature-dependent magnetization (MT), the samples were cooled down to 5 K under a zero field, then a 15 kOe magnetic field was applied and the magnetization data was collected during the warming up process. The transport properties of MnSi films were investigated by a Lake Shore Hall measurement system. Magnetic-field dependent resistance was measured between 5 and 300 K using the van der Pauw geometry. The magnetic field was applied along the sample surface plane (in-plane).

Table 5.1. The parameters of the samples.

Sample no.	substrate	Thickness of regrown layer (nm)	Flash energy density (J/cm^2)	Anneal surface	The ratio of MnSi$_{1.7}$ (%)
A	Si(100)	14	114.97	Mn surface	80.5
B	Si(100)	20	114.97	Mn surface	59.5
C	Si(100)	30	139.5	Si surface	66
D	Si(100)	40	114.97	Mn surface	65
E	Si(100)	60	134.56	Si surface	84.3
F	Si(100)	60	139.5	Si surface	76.7
G	Si(100)	60	139.5	Mn surface	56.6
H	Si(111)	30	139.5	Si surface	85.1
I	Si(111)	40	110.12	Mn surface	67
J	Si(111)	40	114.97	Mn surface	52.5
K	Si(111)	60	139.5	Mn surface	73
L	Si(111)	60	139.5	Si surface	0

5.3 Results

Figure 5.1 (a) shows the XRD pattern of sample G with a 60 nm MnSi film grown on a Si(100) substrate. The (200) and (400) diffraction peaks of the Si substrate are at 33.05° and 69.2°, respectively. The MnSi (210) and (211) Bragg peaks are observed at 44.6° and 49.2°, respectively. According to the powder PDF card (n. 01-081-0484) [178], the MnSi (210) peak is the strongest. Taking into account the intensity ratio between different peaks, the MnSi film grown on Si(100) exhibits a polycrystalline nature. Furthermore, the MnSi$_{1.7}$ (104) peak appears at 26.03°. These two phases (MnSi and MnSi$_{1.7}$) often coexist [105, 172]. The XRD pattern of sample L with a MnSi film grown on a Si(111) substrate is shown in Figure 5.1 (b). The (111) and (222) diffraction peaks of the Si substrate are at 28.4° and 58.9°, respectively. The MnSi(111) and (222) Bragg peaks are observed at 34.2° and 72.4°, respectively. The MnSi(210) peak is also observed at 44.6°, but with much weaker intensity. Considering the intensity ratio between different peaks, the MnSi phase in this sample is highly (111) textured. Within the detection limit, there is no visible peak, which can be assigned to the MnSi$_{1.7}$. From our measurements for other samples with different thickness (not shown), we have found the following: (1) MnSi films on Si(100) are always polycrystalline with the co-existence of the MnSi$_{1.7}$ second phase; (2) MnSi films on Si(111) are highly (111) textured and we can obtain either pure MnSi phase or the mixture of two phases with different concentration ratios.

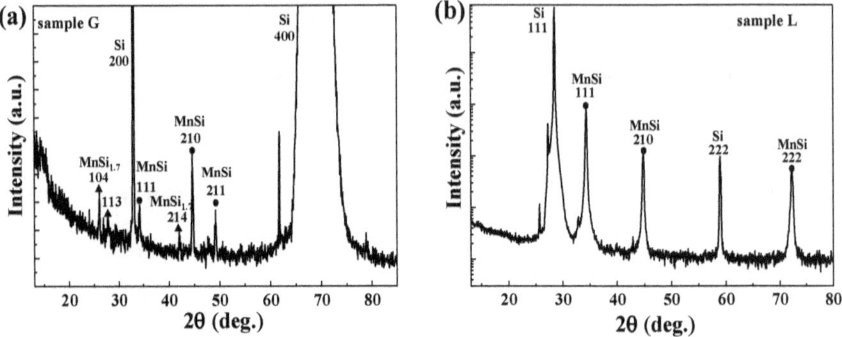

Figure 5.1. (a) XRD pattern of a 60 nm MnSi film on Si(100) by FLA. B20-MnSi and MnSi$_{1.7}$ phases coexist. (b) XRD pattern of a 60 nm MnSi film on Si(111) by FLA, and in this sample B20-type MnSi is the single phase. The insert table in (b) shows the relative intensity ratio of different diffraction planes. Ref. 23 shows the (210) and (211) planes should be the two strongest peaks. Sample G shows consistent with Ref. 23, indicating a polycrystalline structure. The (111) plane of sample L is the strongest peak, meaning the (111)-textured of this sample.

As shown exemplarily in Figure 5.2 (a) and (b) for samples G and L with 60 nm MnSi films grown on Si(100) and (111) substrates, the in-plane magnetic loops at 5 K show a multi-hysteresis feature that occurs only when the magnetic states change, being consistent with the other MnSi MH curves with skyrmions [106, 193]. The saturation magnetization for sample G is around 75 emu/cm^3, which is lower than the value for bulk B20 MnSi due to the co-existence of MnSi$_{1.7}$ parasitic phase. The saturated field for this sample is around 9 kOe. However, the single-phase sample L in Figure 5.2 (b) shows a higher saturation magnetization of 165 emu/cm^3, which is close to that for bulk MnSi. This sample also has a higher saturated field of 13 kOe. The mixture with MnSi$_{1.7}$ does not affect this multi-hysteresis behaviour, indicating that the skyrmion formation only depends on the existence of MnSi phase.

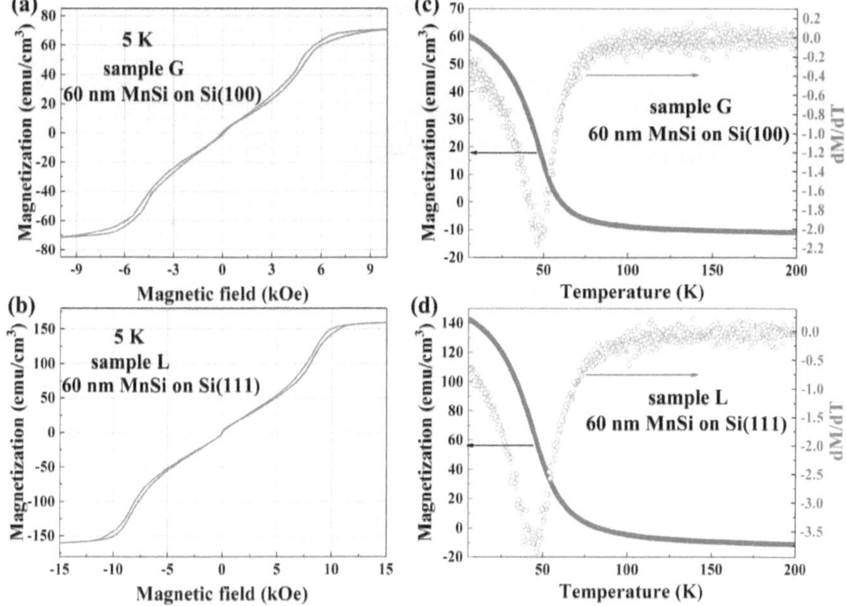

Figure 5.2. In-plane MH curves recorded at 5 K for sample G and L with a 60 nm MnSi films on (a) Si(100) substrates and (b) on Si(111) substrates, respectively. The easy axis and multi-hysteresis are stabilized in-plane. Temperature-dependent in-plane saturation magnetization (solid symbol) and the calculated dM/dT (open symbol) for the samples G (c) and L (d). The valley of dM/dT indicates the Curie temperature.

The Curie temperature of magnetic materials can be measured by different methods, such as temperature dependent magnetization, resistance or heat capacity [115, 194, 195]. From the magnetization, it can be determined by the temperature-dependent remanence [196] and by calculating dM/dT for the temperature-dependent saturation magnetization [197]. T_C can also be determined by measuring the temperature-dependent resistance. In such a case the first derivative dR/dT shows a peak around the critical temperature [198]. At the same time, the temperature-dependent magnetoresistance also shows a peak around T_C [199]. Figure 5.2 (c) and (d) show the in-plane magnetization of samples G and L as a function of temperature under a magnetic field of 15 kOe, which is above the saturation field. The calculated dM/dT curves are shown as open symbols. The minimum of the curves is found at the same temperature for films on Si(100) and (111) substrates, which is defined as the Curie temperature. We also estimated the T_C from electrical measurements (not shown). Both methods, i.e. temperature-dependent magnetization and resistance result in a similar value of T_C at around 43 K.

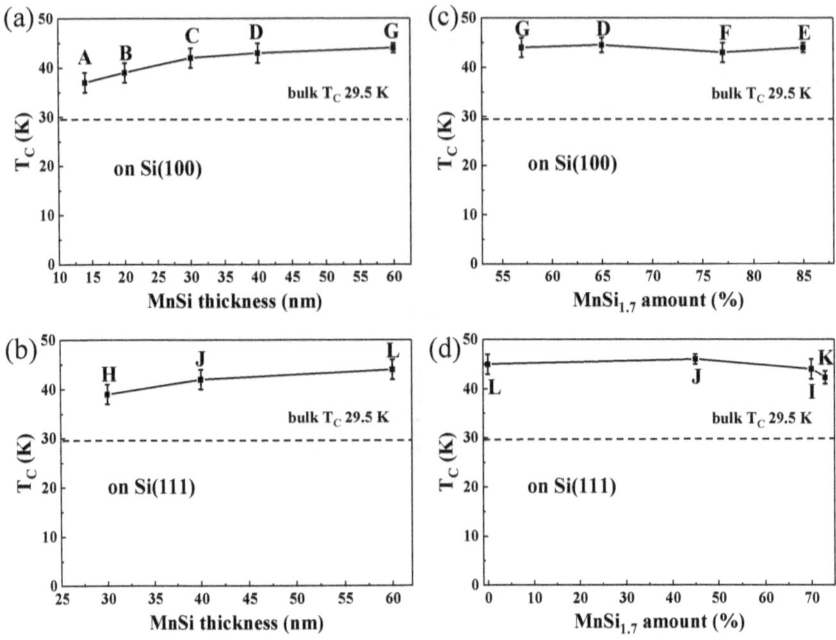

Figure 5.3. Curie temperature as a function of the thickness of MnSi films on (a) Si(100) substrates and Si(111) substrates (b). Curie temperature as a function of the content of MnSi$_{1.7}$ phase on (c) Si(100) substrates and Si(111) substrates (d). All these samples have approximately the same Curie temperature of 43 K, which is larger than that of bulk MnSi. The sample ID is indicated in the figures around the corresponding data point.

The T_C values are plotted in Figure 5.3 (a) and (b) as a function of MnSi film thickness. The T_C slightly increases to 43 K with increasing thickness of the MnSi films on Si(100) or (111) substrates. When the thickness is above 30 nm, T_C saturates at around 43 K. Compared with bulk MnSi, it is increased by 45% and remains stable at 43 K. For thinner samples, T_C is a little bit lower, consistent with Ref. [106]. Although these samples have different crystal orientations or textures on Si(100) or (111) substrates, their T_C is almost the same, indicating that the T_C of MnSi films is not related to the crystal orientation or texture.

In order to check if the presence of MnSi$_{1.7}$ has influence on the T_C of MnSi films [97], we tried to find a correlation between T_C and the content of MnSi$_{1.7}$. The content of MnSi$_{1.7}$ in percent is estimated from the saturation magnetization. Since MnSi$_{1.7}$ is a weak itinerant magnet with negligible magnetization of 0.012 μ_B/Mn [82], we compare the saturation magnetization of our films with that of bulk MnSi. Then the amount of MnSi is determined and we assume that the rest of Mn form $MnSi_{1.7}$. As shown in Figure 5.3 (c, d), the T_C always remains around 43 K with increasing amount of the $MnSi_{1.7}$ phase on Si(100) or (111) substrates. Here, the film thickness is fixed at around 40 and 60 nm and the amount of $MnSi_{1.7}$ is varied since we changed the FLA energy slightly. Therefore, T_C of MnSi films is not determined by its crystalline orientation, texture or the mixture with $MnSi_{1.7}$. In the next section, we investigate the dependence on strain.

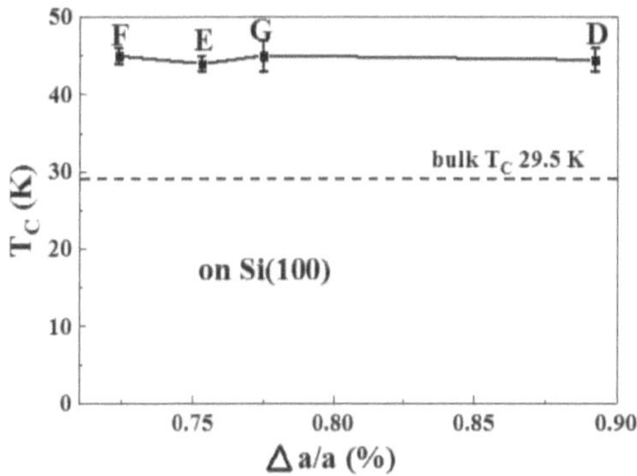

Figure 5.4. Curie temperature as a function of the change of lattice constant 'a' on Si(100) substrates. All these sample have approximately the same Curie temperature of 43 K, which is larger than bulk MnSi. The sample ID is indicated in the figures around the corresponding data point.

Thin films on substrates usually are in a "stressed" state. The strain of thin films can be calculated by the variation of lattice constants. The change of the lattice constant 'a' and 'c' is generally regarded as in-plane and out-of-plane strain, respectively [200], in the case of epitaxy. The change of the lattice constant can be obtained by XRD measurements. The lattice spacing d can be calculated from equation (18) of Bragg's law [201]. θ is the Bragg peak of the specific crystal plane, which can be expressed as the Miller index (hkl). λ is the wavelength of the incident X-ray. n is the diffraction order and is a positive integer.

$$2dsin\theta = n\lambda \tag{18}$$

As shown in equation (19), the lattice constant 'a' for a strained cubic crystal can be obtained.

$$\frac{1}{d^2} = \frac{h^2+k^2+l^2}{a^2} \tag{19}$$

To match the substrates, the lattice of fully strained epitaxial films is locked on the substrates.

Since our films are polycrystalline in general, it is reasonable to calculate the lattice constant like for polycrystalline materials [202]. For the thin film grown on Si(100) substrates, MnSi is fully polycrystalline because the XRD intensity ratios for different diffraction peaks are the same as for the powder sample. Thus, we calculate the change of average lattice constants by equation (19).

Figure 5.4 shows the Curie temperature dependent on the variation of lattice constants 'a' for the MnSi films on Si(100) substrates. These MnSi films have a thickness of 40 or 60 nm. The lattice constants vary due to the different flash lamp annealing energy. As shown in Figure 5.4, the Curie temperatures of all MnSi films stay around 43 K. In the case of growing on Si(100), the lattice constant a is larger than the bulk values. However, for all cases, the T_C of MnSi films is around 43 K and is not determined by the change of the lattice constant 'a', which is somewhat equivalent to the in-plane strain or out-of-plane strain, respectively. The Curie temperature dependency on the volume variation of MnSi films is shown in Figure 5.5. Obviously, despite the lattice cell expansion or contraction (applying hydrostatic press), MnSi films have a T_C around 43 K.

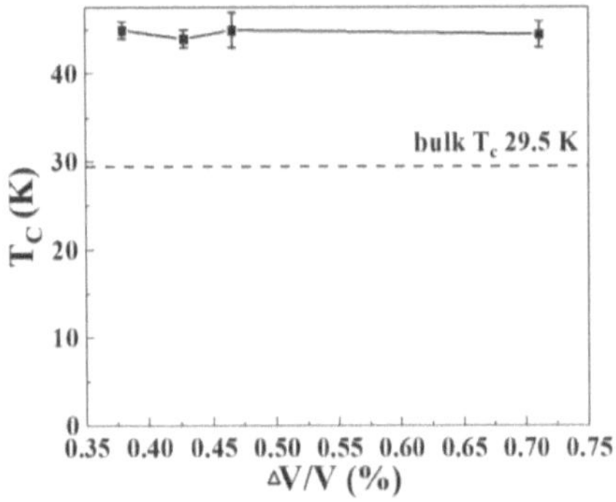

Figure 5.5 The Curie temperature vs. the change of the cell volume of MnSi films on Si(100) substrates. The Curie temperature stays around 43 K with the variety of cell volume. The sample ID is indicated in the figures around the corresponding data point.

From the experimental results shown above, the origin of the increased T_C in MnSi films remains still elusive. We found that, being independent of the crystallinity, the substrate orientation, the lattice strain and the cell volume, T_C is always around 43 K when the thickness is above a threshold value. By detailed investigations using EXAFS, Figueroa *et al.* suggests the interface maybe the reason for the improved Curie temperature. They also suspected that the shifted Si positions within the unit cell could have an indirect effect on the magnetic ordering of the Mn atoms, e.g., via the crystal field. However, they have not detected skyrmions in their films. In our thick films, both magnetic skyrmions and an enhanced Curie temperature are detected [142]. Therefore, the understanding of the magnetic properties of MnSi films is far from satisfactory. Instead, a subtle variation in the microstructure may play an important role. Both theoretical work by Yabuuchi *et al.* [203] and Men'shov *et al.* [204] points to the off-stoichiometry induced spin fluctuations above the intrinsic Curie temperature in MnSi alloys. In experiments, Rylkov *et al.* have found that in Si-Mn films ($MnSi_{1.7}$ and MnSi) with a slight excess of Mn (around 2-5% from their stoichiometry) the Curie temperature increases [205]. However, the skyrmion formation was not discussed in those papers. Very recently, Balasubramanian *et al.* reported ferromagnetic order in B20-CoSi [197]. Perfectly stoichiometric CoSi does not exhibit any kind of magnetic order. $B20\text{-}Co_{1+x}Si_{1-x}$ with excess Co was obtained by nonequilibrium processing. They found that the alloys are magnetically ordered above a critical excess-Co content (x=0.028). Their density functional theory calculation shows that the onset of the zero-temperature magnetism has the character of a magnetic quantum-phase transition. Nevertheless, the inevitable subtle deviation in stoichiometry and point defects in MnSi films seem to play a non-negeligible role in their magnetic properties. Indeed, it is challenging to quantitatively characterize the subtle amount of off-stoichiometry and point defects. To tackle this problem, well-controlled growth of MnSi films and sensitive characterization are pre-requisites. For the latter, positron annihilation spectroscopy may provide more information about the defects [206].

5.4 Conclusion

In summary, in this paper we try to understand the puzzling Curie temperature widely reported in MnSi films. We have prepared MnSi films with a large variation regarding their thickness, crystallinity, strain and phase separation by a fast solid-state reaction though millisecond flash lamp annealing. Particularly, polycrystalline MnSi films on Si(100) and textured MnSi films on Si(111), both with different mixture ratio with $MnSi_{1.7}$ have been grown and systematically characterized. Surprisingly, all obtained MnSi films exhibit a high Curie temperature at around 43 K when the thickness is above a threshold value. The skyrmion phase has also been detected in these films. However, we find no correlation between the increased Curie temperate and the film thickness (above 30 nm), strain, lattice volume or the mixture with $MnSi_{1.7}$. Our work has not provided a conclusive picture for this question, but is rather calling a revisit, especially to the effect by the interface, stoichiometry and other point defects. Further studies are essential to understand the B20 transition-metal silicide/germanides films and therefore to utilize them for spintronic applications.

6. Summary and outlook

6.1 Summary

The aim of the current book was to investigate the preparation of MnSi film on Si substrates. The preparation process includes room temperature sputtering Mn films with different thicknesses and flash-lamp annealing with different energy density (annealing temperature). Systematic investigations on their structural, electrical, magnetic, and magneto-transport properties were performed. The key findings are summarized below:

Thin films with the B20-MnSi phase on Si(100) substrates were fabricated for the first time. They exhibit magnetic skyrmion behaviour. In comparison with Si(111) substrates, Si(100) substrates are more preferred from the practical application point of view. The nucleation of B20-MnSi on Si(100) is believed to be triggered by the fast solid-state phase reaction between Mn and Si via ms-range flash-lamp annealing. Compared with the corresponding bulk material, our films show an increased Curie temperature of around 43 K. The magnetic and transport measurements reveal that skyrmions in B20-MnSi on Si(100) made by sub-seconds solid-state reaction are stable within much broader field and temperature windows than bulk MnSi. The parasitic $MnSi_{1.7}$ phase can be further minimized or eliminated by optimizing the annealing conditions, the quality of the deposited Mn film, and its interface with the Si substrate. Our work demonstrates a promising route for the fabrication of B20-type transition metal silicides for integrated and/or hybrid spintronic applications on Si(100) wafers, which are more preferable for industry applications.

The growth of MnSi films on Si(111) substrates has been widely realized by solid phase epitaxy or molecular beam epitaxy since the lattice mismatch and symmetry fit better. One problem is the parasitic $MnSi_{1.7}$ phase. By controlling the reaction parameters using strongly non-equilibrium flash lamp annealing, we have achieved full control over the phase formation of Mn-silicides in thin films from single-phase B20-MnSi or $MnSi_{1.7}$ to mixed phases. The obtained films are highly textured and reveal sharp interfaces to the Si substrate. The obtained B20-MnSi films exhibits a high Curie temperature of 41 K. The skyrmion phase can be stabilized over broad temperature and magnetic field ranges. We propose flash-lamp-annealing-induced transient reaction as a general approach for phase separation in transition-metal silicides and germanides and for growth of B20-type films with enhanced topological stability.

By comparing the magnetic properties of MnSi films grown on both Si(111) and Si(100) substrates by ourselves and by others in literature, we found one common feature. It is the increased Curie temperature of around 41-43 K for all MnSi films. It is much higher than 29.5 K for bulk MnSi. We try to understand the puzzling Curie temperature widely reported in MnSi films. We have prepared MnSi films with a large variation regarding their thickness, crystallinity, strain and phase separation. Particularly, polycrystalline MnSi films on Si(100) and textured MnSi films on Si(111), both with different mixture ratio with $MnSi_{1.7}$ have been grown and systematically characterized. Surprisingly, all obtained MnSi films exhibit a high Curie temperature at around 43 K. The skyrmion phase has also been detected in these films. However, we find no correlation between the increased Curie temperate and the film thickness, strain, lattice volume or the mixture with $MnSi_{1.7}$. Our work has not provided a conclusive picture for this question, but is rather calling a revisit, especially to the effect by the interface, stoichiometry and point defects. Further studies are essential to understand the B20 transition-metal silicide/germanides films and therefore to utilize them for spintronic applications.

6.2 Outlook

Based on my experience in the past 3 years, I have the following suggestions for future work.

6.2.1 Film thickness effect on formation of (111)-textured B20-MnSi

I have started with this investigation. 15-60 nm thick Mn films were firstly deposited on Si(111) wafers by DC magnetron sputtering. The thickness of the formed Mn-Si films is from around 30 nm to 120 nm, i.e. around the twice of the deposited Mn films as checked by transmission electron microscopy. Afterwards, the flash-lamp annealing was employed to realize a fast solid-state reaction between Mn and Si. The annealing parameters were chosen according to the results in Chapter 4.

Figure 6.1 shows the XRD patterns of the FLA-treated samples on Si(111) substrates. To better understand the thickness dependent of B20-MnSi formation, each sample from different thicknesses is chosen from the optimized annealing condition. When the thickness is thicker than 15 nm, MnSi and $MnSi_{1.7}$ phases are detected. According to the powder PDF card (01-081-0484 and 04-005-9870), the MnSi (210) and $MnSi_{1.7}$ (214) planes should be the strongest diffraction, while the MnSi (111) and $MnSi_{1.7}$ (200) planes are the strongest peak of these two phase. This indicates that the MnSi and $MnSi_{1.7}$ phase in these samples are textured on Si(111) substrates. Due to the lower lattice mismatch of 1.8% and 3.1% for $MnSi_{1.7}$ and MnSi with

Si(111) substrate, respectively, the (200) plane of MnSi$_{1.7}$ and the (111) plane of MnSi turns out to be parallel to Si(111) plane. When the thickness of deposited Mn is around 30 nm, single B20-MnSi phase is obtained. However, the MnSi$_{1.7}$ phase comes back with increasing the thickness.

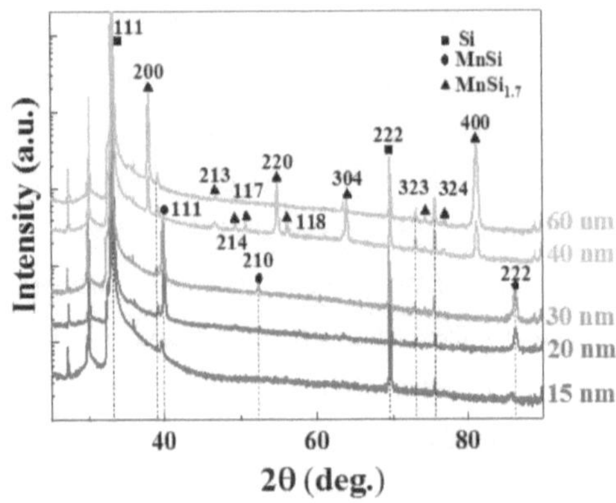

Figure 6.1. The XRD patterns of the regrown layers annealed from various thicknesses of Mn films on Si(111) substrates. B20-MnSi with (111) texture can be formed as the deposited Mn varies from 20 to 30 nm.

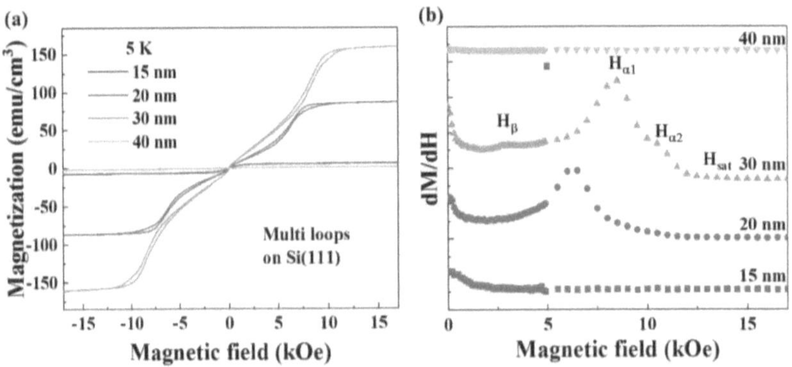

Figure 6.2. (a) In-plane MH curves of the samples annealed from different Mn thicknesses on Si(111) substrates measured at 5 K. The samples annealed from 20- and 30-nm-thick Mn have multi-hysteresis, indicating transitions between different magnetic phases. (b) The calculated dM/dH from the the MH curves from (a). Four critical fields are identified and labelled as H_β, $H_{\alpha 1}$, $H_{\alpha 2}$, and H_{sat}. $H_{\alpha 1}$ and $H_{\alpha 2}$ bound the region of stable elliptic skyrmion. H_β and H_{sat} are the critical fields for the start of helical to skyrmion and the concial to ferromagnetic state. For

the sample annealed from 15- and 40-nm-thick Mn, there is no peak, indicating no magnetic phase transition since they only contain $MnSi_{1.7}$.

Figure 6.2 (a) shows the MH curves of FLA-treated samples from different thicknesses of deposited Mn on Si(111) substrates. The samples annealed from 20 and 30 nm Mn show an obvious multi-hysteresis, indicating the transition between different magnetic structures, consistent with other B20-MnSi films. It is reported that this multi-hysteresis only occurs when the thickness is over 18 nm, which is the theoretical size of skymions in B20-MnSi. Only when the thickness of film is thicker than 18 nm, this skyrmion behaviour can be detected. The sample annealed from 30 nm Mn has the strongest magnetization among all these samples, which is close to the saturation magnetization of reported B20-MnSi compound. With increasing the thickness of deposited Mn to 40 nm, there is also no clear magnetic hysteresis since $MnSi_{1.7}$ dominates the crystalline phase. The optimized annealing condition for 30 nm Mn does not work for thicker films.

We also proved the magnetic skyrmions in samples with different thicknesses. The calculated dM/dH is shown in Figure 6.2 (b). At the beginning, all these curves monotonically decline with increasing the magnetic field. For the samples annealed from 15 and 40 nm Mn, there is no special feature since $MnSi_{1.7}$ is the dominant phase. On the contrary, the samples annealed from 20 and 30 nm Mn have a peak H_β around 3 kOe, which is the transition from helical to skyrmion. By this first order transition, the magnetic state starts to change into topological skyrmion state gradually. With increasing the magnetic field, another peak $H_{\alpha 1}$ comes out, indicating the totally different spin configurations, where all the helical structure transforms into skyrmions. The peak $H_{\alpha 2}$ is the position where the metastable skyrmions begin to change into conical structure. Afterwards, all magnetic structure transforms into ferromagnetic state at the beginning of horizontal curve by a second order transition. The magnetic and transport measurement for the sample annealed from 30 nm Mn on Si(111) substrates can be found in Chapter 4. It is shown that this sample has an extended skyrmion window.

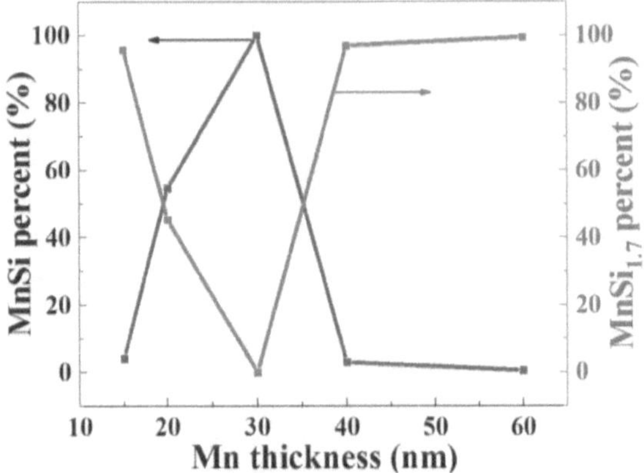

Figure 6.3. The phase ratios of MnSi and MnSi$_{1.7}$ of regrown layers on Si(111) as a function of the thickness of deposited Mn films. The B20-MnSi is preferably formed when Mn thickness is at 20-30 nm. Thicker or thinner than this value results in the MnSi$_{1.7}$ phase.

The ratio of B20-MnSi and the parasitic MnSi$_{1.7}$ phase dependent on thickness can be calculated through the saturated magnetization and is shown in Figure 6.3. As shown in Figure 6.3, the content of B20-MnSi increases with increasing the thickness of deposited Mn. When the deposited Mn has a thickness over 20 nm, the ratio of MnSi increase to more than 50%. The coexistence of these two phases is observed by XRD and TEM. For a film annealed from 30 nm thick Mn, there is single phase MnSi. With further increasing the thickness of Mn films, MnSi$_{1.7}$ starts to become the dominant phase.

These preliminary results imply that the optimized annealing condition to obtain single MnSi phase should also depend on the thickness of Mn films. This can be understandable since the temperature gradient can be a little bit different during flash lamp annealing when the thickness of metal Mn is changed. We propose an empirical explanation as the follows.

94

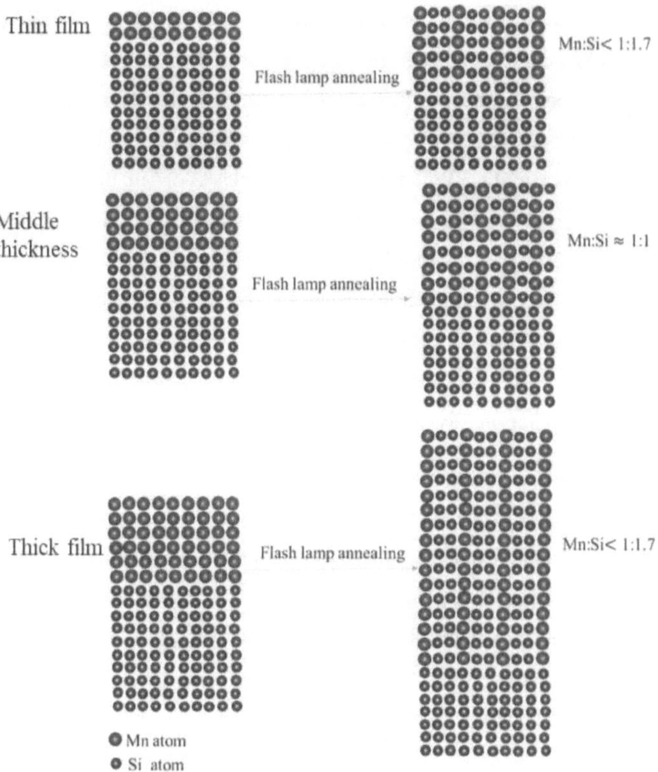

Figure 6.4. The schematic for the diffusion and nucleation process. When the film is very thin, the excessive diffusion of Si makes the composition ratio of Si and Mn larger than 1:1, which can not support the formation of B20-MnSi. When the film is thick, the limited thermal conductivity of Mn 7.7 W/(mK) decreases the reaction temperature, which will not favour the nucleation of MnSi, but MnSi$_{1.7}$.

The schematic for the atom diffusion and nucleation processes are shown in Figure 6.4. It is well known that composition and reaction temperature are the most important parameters to affect the nucleation process. In each flash annealing process, the diffusion of Si atom is excessive in thin films or thick films, which can be confirmed by future RBS measurements. The excessive Si can exist either between the recrystallized MnSi crystallites or through the layer regrown to the surface, which can be confirmed/checked by TEM experiments. When the film is very thin, like 15 nm, the diffusion of Si can dilute the composition ratio of Mn and Si, which is far away from 1:1. Thus it is easy to form the parasitic MnSi$_{1.7}$ phase due to the excessive diffusion of Si. Even in this case, Si still diffuses through the $MnSi_{1.7}$ layer to the surface. The disorder composition and larger mismatch make the formation of MnSi very hard.

It is also well known that Mn has a very small thermal conductivity of 7.7 W/(mK) compared with Si of around 150 W/(mK) [45]. The thermal conductivity can be even smaller due to the nm-size of Mn films. With increasing the thickness of Mn, the reaction temperature of the crystalline front will decrease due to the finite thermal conductivity. The low-temperature phase of the parasitic $MnSi_{1.7}$ will prefer to nucleate at this case.

As a short summary, we still do not understand the phase formation (MnSi vs. $MnSi_{1.7}$) depending on Mn thickness. More experimental works are needed to check if we can prepare thinner (< 40 nm) or thicker (> 60 nm) films with single phase MnSi.

6.2.2 $MnSi_{1.7}$% influence on Skyrmion stability

It is known that the parasitic $MnSi_{1.7}$ phase is much more easily formed together with MnSi. It is not clear how the parasitic $MnSi_{1.7}$ phase affects the magnetic properties of MnSi. In most papers, they authors think that this parasitic phase can be negligible. In Chapter 5, we also found that the $MnSi_{1.7}$ phase has no effect on the Curie temperature of MnSi. However, $MnSi_{1.7}$ might play an important role in the induced anisotropy, which adjusts the required magnetic field for stabilizing different magnetic phases.

I have done some preliminary measurements regarding this question. Figure 6.5 (a) and (c) show the transition fields $H_{\alpha1}$, $H_{\alpha2}$ and H_{sat} marked by the arrows of the sample with 60 nm regrown layers on Si(111) substrates with different ratios of $MnSi_{1.7}$. Values for positive and negative magnetic fields are denoted by the superscripts "+" and "−", respectively. Below the lower critical field $H_{\alpha1}$, the helimagnetic phase is stable. With increasing the magnetic field, the helimagnetic phase transforms into the Skyrmion state via a first order phase transition. Above the upper critical field $H_{\alpha2}$, the system transforms into the conical phase from the Skyrmion state. Beyond the field H_{sat} at the shoulder in Figure 6.5 (a) and (c) (or the kink), the system finally transfers into the field-polarized ferromagnetic state (FM). Helimagnetic phase, Skyrmion and conical phase appear in turn with the increase of the external magnetic field. With increased temperature, the critical fields shifted to lower magnetic field due to the decreased anisotropy.

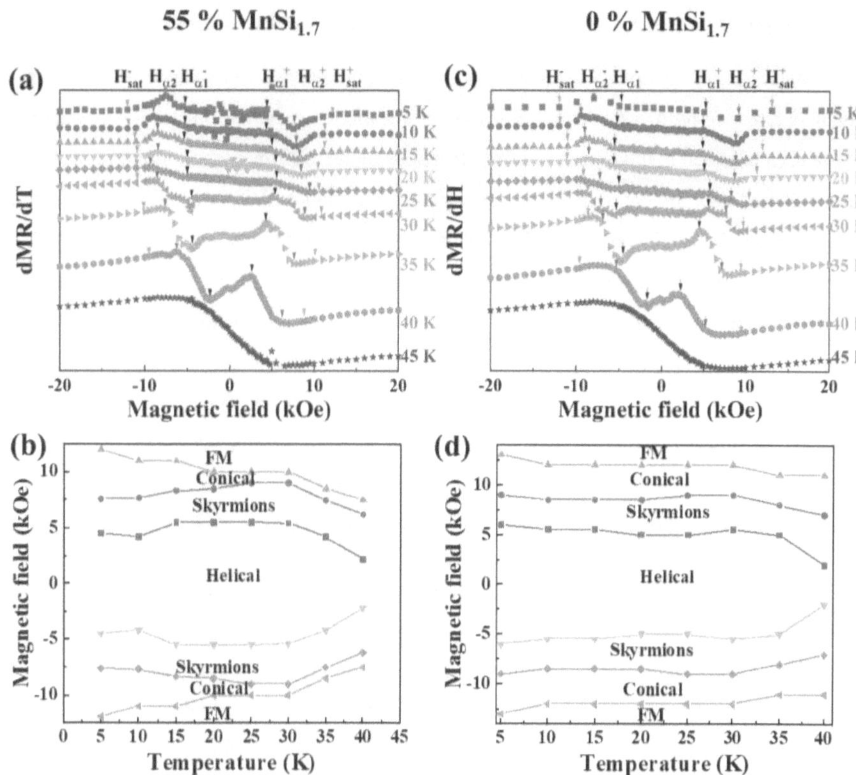

Figure 6.5. Calculated dR/dH of the annealed samples with 55% (a) and 0% (c) MnSi$_{1.7}$ from field-dependent magnetoresistance. The black, red and green arrows stand the position of H$_{\alpha 1}$, H$_{\alpha 2}$ and H$_{sat}$, respectively. Magnetic phase diagram of the annealed samples with 55% (b) and 0% (d) MnSi$_{1.7}$ under in-plane magnetic field inferred from MR data.

Combining the derived dMR/dT, magnetic phase diagrams for these two samples were constructed and shown in Figure 6.5 (b) and (d), in which the solid symbols are from the critical fields Figure 6.5 (a) and (c). In Figure 6.5(b) and (d), the Skyrmion is stabilized over a broad region in the T−H plane, spanning from the T$_C$ down to 5 K for both samples. However, the field range for the skyrmion phase is 4-7 kOe for the film containing MnSi$_{1.7}$ and 6-9 kOe for the film containing no MnSi$_{1.7}$. This film containing MnSi$_{1.7}$ seems to have a decreased induced anisotropy. Of course, this cannot be concluded since there could be other variations that can lead the same effect. In the future, more measurements should be done for samples with different amount of MnSi$_{1.7}$. One also can think about whether the skyrmion phase in MnSi can be better tuned by intentionally including some MnSi$_{1.7}$.

6.2.3 Preparation of other transition-metal monosilicides and germanides

As we have mentioned, B20-type trainsition-metal monosilicides and germanides have many interesting properties. However, as to our knowledge, there are only few materials that have been prepared in the thin film form. Most of available B20 monogermanides require high pressure and high temperature sintering condition. We suggest that the method developed in this book can be used to prepare other trantition-metal monosilicide and germanide films on Si(111) and Ge(111) substrates, respectively. As already demonstrated for MnSi, the growth of B20-type transition metal monosilicides (or germanides) on Si(111) [or Ge(111)] is feasible. The symmetry fits well: Along the [111] crystal axis, both films and substrates have six-fold symmetry. The lattice mismatch between B20-type transition metal monosilicides / germanides and the Si(111)/Ge(111) substrates is relatively small if the film is rotated by 30° with the crystallographic orientation of B20[1-10] ‖ Si/Ge[11-2]. The lattice mismatch can be calculated by $[a_{B20}-a_{Si/Ge}\cos(30^{\circ})]/a_{B20}$ (a is the lattice parameter).

In Table 6.1, we list the lattice mismatch between several B20-type films and Si/Ge. As one can see from the table, this lattice mismatch induces an in-plane tensile strain in the films for 3d transition metals. For 3d transition metal monosilicides, the lattice parameter decreases with increasing the atomic number of the transition metal, consequently the tensile strain will increase. However, the lattice parameter generally increases for 4d transition metal monosilicides and therefore induces a compressive strain when growing on Si(111) substrates. However, there is no experimental report yet as to our best knowledge. We can extend the preparation approach shown in this book to prepare other films.

Table 6.1: Lattice mismatch between reprehensive B20-type silicides (germanides) and the Si (Ge) substrates.

Thin films	MnSi /Si(111)	FeSi /Si(111)	CoSi /Si(111)	RhSi /Si(111)	CoGe /Ge(111)	FeGe /Ge(111)	MnGe /Ge(111)	RhGe /Ge(111)
Lattice mismatch	-3.2%	-4.7%	-5.8%	-0.6%	-5.6%	-4.2%	-2.1%	-0.8%

Acknowledgement

Time goes fast and I have been Dresden for almost 4 years. Each time when I read the acknowledgement of other doctoral book, or took part in a defense, as well as after solving a problem during my research, I was imaging how I would express my gratitude when my doctoral book will be finished in the future. Even though this process is not as easy as one might think when that day actually comes. If there is a timeline, I would first thank the China Scholarship Council (CSC) for their financial support, otherwise I will not be able to continue and complete my doctoral studies in Germany. I sincerely thank Prof. Dr. Kornelius Nielsch and Prof. Dr. Manfred Helm for the supervision of my PhD work at IFW Dresden and HZDR in association with Faculty of Mechanical Science and Engineering at Technische Universität Dresden (TU Dresden).

I would like to thank my direct supervisor Dr. Shengqiang Zhou for giving me the opportunity to work at the Institute of Ion Beam Physics and Materials Research at HZDR. Here, I have great freedom to try out different scientific ideas while his theoretical knowledge and practical experience drive my research work effectively and successfully. He always listens to me and gives me valuable suggestions and encouragement. Due to his positive and approachable personality, which is humble and polite in dealing with people, I never felt like I was talking to a superior when discussing with him. Honestly, he is more of a friend than an advisor.

Here I also want to thank my former direct supervisor Dr. Simon Pauly for giving me the opportunity to study in Germany and work at the Institute for Complex Materials (IKM), IFW Dresden. He also taught me independence and initiatives as a PHD student. This work would never be a success without the substantial support from my colleagues at HZDR and IFW. I convey great thanks to Mr. Viktor Begeza and Dr. Christoph Folgner for Mn deposition. I am also thankful for Dr. Yuelei Zhao and Dr. Ruslan Salikhov for the deposition of Co film. Especially, Dr. Slawomir Prucnal who is responsible for the Raman and Flash-lamp annealing devices, whenever I faced problems in the work, he always gave me instant help and discussion. Dr. Lars Rebohle also helped me a lot about the temperature calculation and basic knowledge of Flash lamp annealing. Prof. Dr. Olav Hellwig is very kind and allow me to use their XRD and sputtering. I also thank Dr. Jörg Grenzer and Mrs. Andrea Scholz for teaching me XRD. Dr. René Hübner is thanked for his TEM measurements and analysis. I also thank our technician Mr. Thomas Schumann and Mrs. Ilona Skorupa for their help.

I owe special gratefulness to my colleagues in building 104, Dr. Mao Wang, Dr. Xu Chi, Dr. Changan Wang, Dr. Lei Cao, Mr. Zhen Shang, Ms. Juanmei Duan, I will never forget the days we worked together and played together during the time when I came to HZDR. I am also extremely thankful to Ms. Yufang Xie, Mr. Tianbing He, Mr. Le Feng, Mr. Yujian Zhou, Mr. Qiongqiong Lu, Mrs. Ling Ding, Mr. Panpan Zhao, Mr. Angelo Fernandes Andreoli Dr. Pramote Thirathipviwat, and Dr. Kosiba Konard for doing PhD and sharing unforgettable PhD experience together from the beginning.

Grateful thanks are due to all my Chinese colleagues and friends in Dresden, Dr. Pei Wang, Dr. Liang Deng, Dr. Ju Wang, Dr. Wei Wei, Dr. Peng Xue, Dr. Lixia Xi, Dr. Ran He, Dr. Jinbo Pang, Dr. Jing Guo, Mrs. Xi Ran, Mr. Chenghe Wang, Mr. Yichao Liu, Ms. Erjuan Guo, Mr. Tianyu Tang, Mr. Tao Huang, Ms. Kenan Zhang and Mr. Peng Wang, who helped me in academic and/or daily life, organized the Chinese food party to relieve my homesickness in the festivals. I am also indebted to the Chinese church of Dresden, where the weekends I spent with them and that provided me an effective way to release stress and keep happy.

I extend my special thanks from the core of my heart to a special person, Ms. Linjie Zheng, who stands by me and gives me unwavering love, support and patience. Thank you for sharing happiness, dividing sorrows and imagining the future with me. Finally, I would like to express my heartfelt gratitude to my parents. I couldn't have come to this end without your constant understanding and support.

(Zichao Li)

References

[1] West A. R. Basic Solid State Chemistry. John Wiley & Sons Ltd.: New York, (1984).

[2] Ahmed A. S., Esser B. D., Rowland J., McComb D. W., Kawakami R. K. Molecular beam epitaxy growth of [CrGe/MnGe/FeGe] superlattices: Toward artificial B20 skyrmion materials with tunable interactions. Journal of Crystal Growth (2017) 467: 38-46.

[3] Pfleiderer C., Julian S. R., Lonzarich G. G. Non-Fermi-liquid nature of the normal state of intinerant-electron ferromagnets. Nature (2001) 414: 427-430.

[4] Jeong T., Pickett W. E. Implications of the B20 crystal structure for the magnetoelectronic structure of MnSi. Physical Review B (2004) 70: 075114.

[5] Mühlbauer S., Binz B., Jonietz F., Pfleiderer C., Rosch A., Neubauer A., Georgii R., Böni P. Skyrmion Lattice in a Chiral Magnet. Science (2009) 323: 915-919.

[6] Lin Y-C., Chen Y., Shailos A., Huang Y. Detection of Spin Polarized Carrier in Silicon Nanowire with Single Crystal MnSi as Magnetic Contacts. Nano Letters (2010) 10: 2281-2287.

[7] Huang S. X., Chien C. L. Extended Skyrmion Phase in Epitaxial FeGe(111)Thin Films. Physical Review Letters (2012) 108: 267201.

[8] Langner M. C., Roy S., Mishra S. K., Lee J. C. T., Shi X. W., Hossain M. A., et al. Coupled Skyrmion Sublattices in Cu_2OSeO_3. Physical Review Letters (2014) 112: 167202.

[9] Huang S. X., Chen F., Kang J., Zang J., Shu G. J., Chou F. C., et al. Unusual magnetoresistance in cubic B20 $Fe_{0.85}Co_{0.15}Si$ chiral magnets. New Journal of Physics (2016) 18: 065010.

[10] Xu B., Fang Z., Sánchez-Martínez M-Á., Venderbos J. W. F., Ni Z., Qiu T., et al. Optical signatures of multifold fermions in the chiral topological semimetal CoSi. Proceedings of the National Academy of Sciences (2020) 117: 27104-27110.

[11] Sanchez D. S., Belopolski I., Cochran T. A., Xu X., Yin J-X., Chang G., et al. Topological chiral crystals with helicoid-arc quantum states. Nature (2019) 567: 500-505.

[12] Rao Z., Li H., Zhang T., Tian S., Li C., Fu B., et al. Observation of unconventional chiral fermions with long Fermi arcs in CoSi. Nature (2019) 567: 496-499.

[13] Miao H., Zhang T. T., Wang L., Meyers D., Said A. H., Wang Y. L., et al. Observation of Double Weyl Phonons in Parity-Breaking FeSi. Physical Review Letters (2018) 121: 035302.

[14] Tang P., Zhou Q., Zhang S-C. Multiple Types of Topological Fermions in Transition Metal Silicides. Physical Review Letters (2017) 119: 206402.

[15] Bergmann Kvon., Kubetzka A., Pietzsch O., Wiesendanger R. Interface-induced chiral domain walls, spin spirals and skyrmions revealed by spin-polarized scanning tunneling microscopy. Journal of Physics: Condensed Matter (2014) 26: 394002.

[16] Soumyanarayanan A., Reyren N., Fert A., Panagopoulos C. Emergent phenomena induced by spin–orbit coupling at surfaces and interfaces. Nature (2016) 539: 509-517.

[17] Yin J.-X., Zhang S. S., Li H., Jiang K., Chang G., Zhang B., et al. Giant and anisotropic many-body spin–orbit tunability in a strongly correlated kagome magnet. Nature (2018) 562: 91-95.

[18] Ran K., Liu Y., Guang Y., Burn D. M., van der Laan G., Hesjedal T., et al. Creation of a Chiral Bobber Lattice in Helimagnet-Multilayer Heterostructures. Physical Review Letters (2021) 126: 017204.

[19] Roessli B., Böni P., Fischer W. E., Endoh Y. Chiral Fluctuations in MnSi above the Curie temperature. Physical Review Letters (2002) 88: 237204.

[20] Wang C., Du H., Zhao X., Jin C., Tian M., Zhang Y., et al. Enhanced Stability of the Magnetic Skyrmion Lattice Phase under a Tilted Magnetic Field in a Two-Dimensional Chiral Magnet. Nano Letters (2017) 17: 2921-2927.

[21] Wang S., Tang J., Wang W., Kong L., Tian M., Du H. Electrical Detection of Magnetic Skyrmions. Journal of Low Temperature Physics (2019) 197: 321-336.

[22] Heinze S., von Bergmann K., Menzel M., Brede J., Kubetzka A., Wiesendanger R., et al. Spontaneous atomic-scale magnetic skyrmion lattice in two dimensions. Nature Physics (2011) 7: 713-718.

[23] Fert A., Cros V., Sampaio J. Skyrmions on the track. Nature Nanotechnology (2013) 8: 152-156.

[24] Zhang X., Zhou Y., Ezawa M. Magnetic bilayer-skyrmions without skyrmion Hall effect. Nature Communications (2016) 7: 10293.

[25] Wiesendanger R. Nanoscale magnetic skyrmions in metallic films and multilayers: a new twist for spintronics. Nature Reviews Materials (2016) 1: 16044.

[26] Fert A., Reyren N., Cros V. Magnetic skyrmions: advances in physics and potential applications. Nature Reviews Materials (2017) 2: 17031.

[27] Skyrm T. H. R. A unified field theory of mesons and baryons. Nuclear Physics (1962) 31: 556-569.

[28] Kanazawa N., Seki S., Tokura Y. Noncentrosymmetric Magnets Hosting Magnetic Skyrmions. Advanced Materials (2017) 29: 1603227.

[29] Göbel B., Mertig I., Tretiakov O. A. Beyond skyrmions: Review and perspectives of alternative magnetic quasiparticles. Physics Reports (2021) 895: 1-28.

[30] Back C., Cros V., Ebert H., Everschor-Sitte K., Fert A., Garst M., et al. The 2020 skyrmionics roadmap. Journal of Physics D: Applied Physics (2020) 53: 363001.

[31] Chen S., Yuan S., Hou Z., Tang Y., Zhang J., Wang T., et al. Recent Progress on Topological Structures in Ferroic Thin Films and Heterostructures. Advanced Materials (2020) 33: 2000857.

[32] Jiang W., Chen G., Liu K., Zang J., te Velthuis S. G. E., Hoffmann A. Skyrmions in magnetic multilayers. Physics Reports (2017) 704: 1-49.

[33] Li J., Tan A., Moon K. W., Doran A., Marcus M. A., Young A. T., et al. Tailoring the topology of an artificial magnetic skyrmion. Nature Communications (2014) 5: 4704.

[34] Zhang X., Xia J., Zhou Y., Liu X., Zhang H., Ezawa M. Skyrmion dynamics in a frustrated ferromagnetic film and current-induced helicity locking-unlocking transition. Nature Communications (2017) 8: 1717.

[35] Zhang X., Cai W., Zhang X., Wang Z., Li Z., Zhang Y., et al. Skyrmions in Magnetic Tunnel Junctions. ACS Applied Materials & Interfaces (2018) 10: 16887-16892.

[36] Meng K-Y., Ahmed A. S., Baćani M., Mandru A-O., Zhao X., Bagués N., et al. Observation of Nanoscale Skyrmions in $SrIrO_3$/$SrRuO_3$ Bilayers. Nano Letters (2019) 19: 3169-3175.

[37] Deng L., Wu H-C., Litvinchuk A. P., Yuan N. F. Q., Lee J-J., Dahal R., et al. Room-temperature skyrmion phase in bulk Cu_2OSeO_3 under high pressures. Proceedings of the National Academy of Sciences (2020) 117: 8783-8787.

[38] Petrović A. P., Raju M., Tee X. Y., Louat A., Maggio-Aprile I., Menezes R. M., et al. Skyrmion-(Anti) Vortex Coupling in a Chiral Magnet-Superconductor Heterostructure. Physical Review Letters (2021) 126: 117205.

[39] Soumyanarayanan A., Raju M., Gonzalez Oyarce A. L., Tan A. K. C., Im M-Y., Petrović A. P., et al. Tunable room-temperature magnetic skyrmions in Ir/Fe/Co/Pt multilayers. Nature Materials (2017) 16: 898-904.

[40] Maccariello D., Legrand W., Reyren N., Garcia K., Bouzehouane K., Collin S., et al. Electrical detection of single magnetic skyrmions in metallic multilayers at room temperature. Nature Nanotechnology (2018) 13: 233-237.

[41] Chen R., Li C., Li Y., Miles J. J., Indiveri G., Furber S., et al. Nanoscale Room-Temperature Multilayer Skyrmionic Synapse for Deep Spiking Neural Networks. Physical Review Applied (2020) 14: 014096.

[42] Yu X. Z., Onose Y., Kanazawa N., Park J. H., Han J. H., Matsui Y., et al. Real-space observation of a two-dimensional skyrmion crystal. Nature (2010) 465: 901-904.

[43] Fujishiro Y., Kanazawa N., Nakajima T., Yu X. Z., Ohishi K., Kawamura Y., et al. Topological transitions among skyrmion- and hedgehog-lattice states in cubic chiral magnets. Nature Communications (2019) 10: 1059.

[44] Yu X. Z., Kanazawa N., Onose Y., Kimoto K., Zhang W. Z., Ishiwata S., et al. Near room-temperature formation of a skyrmion crystal in thin-films of the helimagnet FeGe. Nature Materials (2010) 10: 106-109.

[45] Tanigaki T., Shibata K., Kanazawa N., Yu X., Onose Y., Park H. S., et al. Real-Space Observation of Short-Period Cubic Lattice of Skyrmions in MnGe. Nano Letters (2015) 15: 5438-5442.

[46] Tokunaga Y., Yu X. Z., White J. S., Rønnow H. M., Morikawa D., Taguchi Y., et al. A new class of chiral materials hosting magnetic skyrmions beyond room temperature. Nature Communications (2015) 6: 7638.

[47] Nayak A. K., Kumar V., Ma T., Werner P., Pippel E., Sahoo R., et al. Magnetic antiskyrmions above room temperature in tetragonal Heusler materials. Nature (2017) 548: 561-566.

[48] Chacon A., Heinen L., Halder M., Bauer A., Simeth W., Mühlbauer S., et al. Observation of two independent skyrmion phases in a chiral magnetic material. Nature Physics (2018) 14: 936-941.

[49] Yu X. Z., Maxim M., Tokunaga Y., Zhang W., Kimoto K., et al. Magnetic stripes and skyrmions with helicity reversals. Proceedings of the National Academy of Sciences (2012) 5: 109 8856-8860.

[50] Yu X. Z., Tokunaga Y., Kaneko Y., Zhang W. Z., Kimoto K., Matsui Y., et al. Biskyrmion states and their current-driven motion in a layered manganite. Nature Communications (2014) 5: 3198.

[51] Wang W., Zhang Y., Xu G., Peng L., Ding B., Wang Y., et al. A Centrosymmetric Hexagonal Magnet with Superstable Biskyrmion Magnetic Nanodomains in a Wide Temperature Range of 100-340 K. Advanced Materials (2016) 28: 6887-6893.

[52] Sun L., Cao R. X., Miao B. F., Feng Z., You B., Wu D., et al. Creating an Artificial Two-Dimensional Skyrmion Crystal by Nanopatterning. Physical Review Letters (2013) 110: 167209.

[53] Fernández-Pacheco A., Vedmedenko E., Ummelen F., Mansell R., Petit D., Cowburn R. P. Symmetry-breaking interlayer Dzyaloshinskii–Moriya interactions in synthetic antiferromagnets. Nature Materials (2019) 18: 679-684.

[54] Han D-S., Lee K., Hanke J-P., Mokrousov Y., Kim K-W., Yoo W., et al. Long-range chiral exchange interaction in synthetic antiferromagnets. Nature Materials (2019) 18: 703-708.

[55] Xu L., Fan J., Sun W., Zhu Y., Hu D., Liu J., et al. Magnetic field-driven 3D-Heisenberg-like phase transition in single crystalline helimagnet FeGe. Applied Physics Letters (2017) 111: 052406.

[56] Pfleiderer C. Surfaces get hairy. Nature Physics (2011) 7: 673-674.

[57] Skjærvø S. H, Marrows C. H., Stamps R. L., Heyderman L. J. Advances in artificial spin ice. Nature Reviews Physics (2019) 2: 13-28.

[58] Menzel D., Engelke J., Reimann T., Süllow S. Enhanced critical fields in MnSi thin films. Journal of the Korean Physical Society (201) 62: 1580-1583.

[59] Kanazawa N., White J. S., Rønnow H. M., Dewhurst C. D., Fujishiro Y., Tsukazaki A., et al. Direct observation of anisotropic magnetic field response of the spin helix in FeGe thin films. Physical Review B (2016) 94: 184432.

[60] Ukleev V., Yamasaki Y., Morikawa D., Karube K., Shibata K., Tokunaga Y., et al. Element-specific soft x-ray spectroscopy, scattering, and imaging studies of the skyrmion-hosting compound $Co_8Zn_8Mn_4$. Physical Review B (2019) 99: 14408.

[61] Wild J., Thomas N. G. M., Pöllath S., Kronseder M., Bauer A., et al. Entropy-limited topological protection of skyrmions. Science Advances (2017) 3: e1701704.

[62] Wang Y., Wang L., Xia J., Lai Z., Tian G., Zhang X., et al. Electric-field-driven non-volatile multi-state switching of individual skyrmions in a multiferroic heterostructure. Nature Communications (2020) 11: 3577.

[63] Samatham S. S., Venkateshwarlu D., Gangrade M., Ganesan V. Magnetotransport studies on polycrystalline MnSi. AIP Conference Proceedings (2012) 983: 1447.

[64] Porter N. A., Gartside J. C., Marrows C. H. Scattering mechanisms in textured FeGe thin films: Magnetoresistance and the anomalous Hall effect. Physical Review B (2014) 90: 024403.

[65] Woo S., Litzius K., Krüger B., Im M-Y., Caretta L., Richter K., et al. Observation of room-temperature magnetic skyrmions and their current-driven dynamics in ultrathin metallic ferromagnets. Nature Materials (2016) 15: 501-506.

[66] Gallagher J. C., Meng K. Y., Brangham J. T., Wang H. L., Esser B. D., McComb D. W., et al. Robust Zero-Field Skyrmion Formation in FeGe Epitaxial Thin Films. Physical Review Letters (2017) 118: 027201.

[67] Pappas C., Lelièvre-Berna E., Falus P., Bentley P. M., Moskvin E., Grigoriev S., et al. Chiral Paramagnetic Skyrmion-like Phase in MnSi. Physical Review Letters (2009) 102: 197202.

[68] Rybakov F. N., Borisov A. B., Blügel S., Kiselev N. S. New Type of Stable Particlelike States in Chiral Magnets. Physical Review Letters (2015) 115: 117201.

[69] Morikawa D., Yamasaki Y., Kanazawa N., Yokouchi T., Tokura Y., et al. Determination of crystallographic chirality of MnSi thin film grown on Si (111) substrate. Physical Review Materials (2020) 4: 014407.

[70] Grigoriev S. V., Chernyshov D., Dyadkin V. A., Dmitriev V., Moskvin E. V., et al. Interplay between crystalline chirality and magnetic structure in $Mn_{1-x}Fe_xSi$. Physical Review B (2010) 81: 012408.

[71] Yang S-H., Naaman R., Paltiel Y., Parkin S. S. P. Chiral spintronics. Nature Reviews Physics (2021) 3: 328-343.

[72] Li Y., Kanazawa N., Yu X. Z., Tsukazaki A., Kawasaki M., Ichikawa M., et al. Robust Formation of Skyrmions and Topological Hall Effect Anomaly in Epitaxial Thin Films of MnSi. Physical Review Letters (2013) 110: 117202.

[73] Yu X., DeGrave J. P., Hara Y., Hara T., Jin S., Tokura Y. Observation of the Magnetic Skyrmion Lattice in a MnSi Nanowire by Lorentz TEM. Nano Letters (2013) 13: 3755-3759.

[74] Wollants P. ternary alloy systems. Springer. 2007.

[75] Reiner M., Bauer A., Leitner M., Gigl T., Anwand W., Butterling M., et al. Positron spectroscopy of point defects in the skyrmion-lattice compound MnSi. Scientific Reports (2016) 6: 29109.

[76] Zhao J. Methods for Phase Diagram Determination. Elsevier Science. (2007).

[77] Okamoto MES H., Mueller E. M. Binary Alloy Phase Diagrams. ASM International. (2016).

[78] Venables J. A., Spiller G. D. T. Nucleation and Growth of Thin Films. Surface Mobilities on Solid Materials (1983) 86: 341-404.

[79] Torrent-Burgués J. The Gibbs energy and the driving force at crystallization from solution. Journal of Crystal Growth (1994) 140: 1-2 107-114.

[80] Cahn J. W., Hilliard J. E. Free Energy of a Nonuniform System. I. Interfacial Free Energy. The Journal of Chemical Physics (1958) 28: 258-267.

[81] Cao Y., Bennett J. C., Dunlap R. A., Obrovac M. N. Li Insertion in Ball Milled Si-Mn Alloys. Journal of The Electrochemical Society (2018) 165: A1734- A1740.

[82] Gottlieb U., Sulpice A., Lambert-Andron B., Laborde O. Magnetic properties of single crystalline Mn_4Si_7. Journal of Alloys and Compounds (2003) 361: 13-18.

[83] Gao Z., Xiong Z., Li J., Lu C., Zhang G., Zeng T., et al. Enhanced thermoelectric performance of higher manganese silicides by shock-induced high-density dislocations. Journal of Materials Chemistry A (2019) 7: 3384-3390.

[84] Okamoto N. L., Koyama T., Kishida K., Tanaka K., Inui H. Crystal structure and thermoelectric properties of chimney–ladder compounds in the Ru_2Si_3–Mn_4Si_7 pseudobinary system. Acta Materialia (2009) 57: 5036-5045.

[85] Liu W-D., Chen Z-G., Zou J. Eco-Friendly Higher Manganese Silicide Thermoelectric Materials: Progress and Future Challenges. Advanced Energy Materials (2018) 8: 1800056.

[86] Chen X., Zhou J., Goodenough J. B., Shi L. Enhanced thermoelectric power factor of Re-substituted higher manganese silicides with small islands of MnSi secondary phase. Journal of Materials Chemistry C (2015) 3: 10500-10508.

[87] Pfleiderer C., Bœuf J., Löhneysen Hv. Stability of antiferromagnetism at high magnetic fields inMn_3Si. Physical Review B (2002) 65: 172404.

[88] Hortamani M., Sandratskii L., Zahn P., Mertig I. Physical origin of the incommensurate spin spiral structure in Mn_3Si. Journal of Applied Physics (2009) 105: 07E506.

[89] Hermann R., Wendrock H., Rodan S., Rößler U. K., Blum C. G. F., Wurmehl S., et al. Single crystal growth of antiferromagnetic Mn_3Si by a two-phase RF floating-zone method. Journal of Crystal Growth (2013) 363: 1-6.

[90] Sürgers C., Fischer G., Winkel P., Löhneysen Hv. Large topological Hall effect in the non-collinear phase of an antiferromagnet. Nature Communications (2014) 5: 3400.

[91] Sürgers C., Kittler W., Wolf T., Löhneysen Hv. Anomalous Hall effect in the noncollinear antiferromagnet Mn_5Si_3. AIP Advances (2016) 6: 055604.

[92] Pamplin B. R. Crystal Growth: International Series on the Science of the Solid State. (1975) 16.

[93] Evers J., Klüfers P., Staudigl R., Stallhofer P. Czochralski's Creative Mistake: A Milestone on the Way to the Gigabit Era. Angewandte Chemie International Edition (2003) 42: 5684-5498.

[94] Keck P., Golay M. Crystallization of silicon from a floating liquid zone. Physical Review, (1953) 89: 1297.

[95] Nii Y., Nakajima T., Kikkawa A., Yamasaki Y., Ohishi K., Suzuki J., et al. Uniaxial stress control of skyrmion phase. Nature Communications (2015) 6: 8539.

[96] Yu X., Kikkawa A., Morikawa D., Shibata K., Tokunaga Y., Taguchi Y., et al. Variation of skyrmion forms and their stability in MnSi thin plates. Physical Review B (2015) 91: 054411.

[97] Sukhanov A. S., Vir P., Heinemann A., Nikitin S. E., Kriegner D., Borrmann H., et al. Giant enhancement of the skyrmion stability in a chemically strained helimagnet. Physical Review B (2019) 100: 180403(R).

[98] Dhital C., DeBeer-Schmitt L., Young D. P., DiTusa J. F. Unpinning the skyrmion lattice in MnSi: Effect of substitutional disorder. Physical Review B (2019) 99: 024428.

[99] Butenko A. B., Leonov A. A., Rößler U. K., Bogdanov A. N. Stabilization of skyrmion textures by uniaxial distortions in noncentrosymmetric cubic helimagnets. Physical Review B (2010) 82: 052403.

[100] Geisler B., Kratzer P., Suzuki T., Lutz T., Costantini G., Kern K. Growth mode and atomic structure of MnSi thin films on Si(111). Physical Review B (2012) 86: 115428.

[101] Wilson M. N., Karhu E. A., Quigley A. S., Rößler U. K., Butenko A. B., Bogdanov A. N., et al. Extended elliptic skyrmion gratings in epitaxial MnSi thin films. Physical Review B (2012) 86: 144420.

[102] Karhu E., Kahwaji S., Monchesky T. L., Parsons C., Robertson M. D., Maunders C. Structure and magnetic properties of MnSi epitaxial thin films. Physical Review B (2010) 82: 184417.

[103] Magnano E., Bondino F., Cepek C., Parmigiani F., Mozzati M. C. Ferromagnetic and ordered MnSi(111) epitaxial layers. Applied Physics Letters (2010) 96: 152503.

[104] Robertson M. D, Karhu E. A., Monchesky T. L. TEM Characterization of Epitaxial MnSi Films Grown on (111) Si Substrates. Microscopy and Microanalysis (2010) 16: 1228-1229.

[105] Karhu E. A., Kahwaji S., Robertson M. D., Fritzsche H., Kirby B. J., Majkrzak C. F., et al. Helical magnetic order in MnSi thin films. Physical Review B (2011) 84: 060404(R).

[106] Karhu E. A., Rößler U. K., Bogdanov A. N., Kahwaji S., Kirby B. J., Fritzsche H., et al. Chiral modulations and reorientation effects in MnSi thin films. Physical Review B (2012) 85: 094429.

[107] Geisler B., Kratzer P. Strain stabilization and thickness dependence of magnetism in epitaxial transition metal monosilicide thin films on Si(111). Physical Review B (2013) 88: 115433.

[108] Figueroa A. I., Zhang S. L., Baker A. A., Chalasani R., Kohn A., Speller S. C., et al. Strain in epitaxial MnSi films on Si(111) in the thick film limit studied by polarization-dependent extended x-ray absorption fine structure. Physical Review B (2016) 94: 174107.

[109] Trabel M., Tarakina N. V., Pohl C., Constantino J. A., Gould C., Brunner K., et al. Twin domains in epitaxial thin MnSi layers on Si(111). Journal of Applied Physics (2017) 121: 245310.

[110] Meynell S. A., Wilson M. N., Krycka K. L., Kirby B. J., Fritzsche H., Monchesky T. L. Neutron study of in-plane skyrmions in MnSi thin films. Physical Review B (2017) 96: 054402.

[111] Wilson M. N., Karhu E. A., Lake D. P., Quigley A. S., Meynell S., Bogdanov A. N., et al. Discrete helicoidal states in chiral magnetic thin films. Physical Review B (2013) 88: 214420.

[112] Yokouchi T., Kanazawa N., Tsukazaki A., Kozuka Y., Kikkawa A., Taguchi Y., et al. Formation of In-plane Skyrmions in Epitaxial MnSi Thin Films as Revealed by Planar Hall Effect. Journal of the Physical Society of Japan (2015) 84: 104708.

[113] Higgins J. M., Ding R., DeGrave J. P., Jin S. Signature of Helimagnetic Ordering in Single-Crystal MnSi Nanowires. Nano Letters (2010) 10: 1605-1610.

[114] Schmitt A. L., Higgins J. M., Szczech J. R., Jin S. Synthesis and applications of metal silicidenanowires. J Mater Chem (2010) 20: 223-235.

[115] Du H., DeGrave J. P., Xue F., Liang D., Ning W., Yang J., et al. Highly Stable Skyrmion State in Helimagnetic MnSi Nanowires. Nano Letters (2014) 14: 2026-2032.

[116] Tang S., Kravchenko I., Yi J., Cao G., Howe J., Mandrus D., et al. Growth of skyrmionic MnSi nanowires on Si: Critical importance of the SiO2 layer. Nano Research (2014) 7: 1788-1796.

[117] Du H., Liang D., Jin C., Kong L., Stolt M. J., Ning W., et al. Electrical probing of field-driven cascading quantized transitions of skyrmion cluster states in MnSi nanowires. Nature Communications (2015) 6: 7637.

[118] Liang D., DeGrave J. P., Stolt M. J., Tokura Y., Jin S. Current-driven dynamics of skyrmions stabilized in MnSi nanowires revealed by topological Hall effect. Nature Communications (2015) 6: 8217.

[119] Okulov I. V., Soldatov I. V., Sarmanova M. F., Kaban I., Gemming T., Edström K., et al. Flash Joule heating for ductilization of metallic glasses. Nature Communications (2015) 6: 7932.

[120] Kosiba K., Scudino S., Kobold R., Kühn U., Greer A. L., Eckert J., et al. Transient nucleation and microstructural design in flash-annealed bulk metallic glasses. Acta Materialia (2017) 127: 416-425.

[121] Orava J., Greer A. L., Gholipour B., Hewak D. W., Smith C. E. Characterization of supercooled liquid $Ge_2Sb_2Te_5$ and its crystallization by ultrafast-heating calorimetry. Nature Materials (2012) 11: 279-283.

[122] Song K. K., Han X. L., Pauly S., Qin Y. S., Kosiba K., Peng C. X., et al. Rapid and partial crystallization to design ductile CuZr-based bulk metallic glass composites. Materials & Design (2018) 139: 132-140.

[123] De Schutter B., De Keyser K., Lavoie C., Detavernier C. Texture in thin film silicides and germanides: A review. Applied Physics Reviews (2016) 3: 031302.

[124] Rebohle L., Prucnal S., Skorupa W. A review of thermal processing in the subsecond range: semiconductors and beyond. Semiconductor Science and Technology (2016) 31: 103001.

[125] Begeza V., Mehner E., Stöcker H., Xie Y., García A., Hübner R., et al. Formation of Thin NiGe Films by Magnetron Sputtering and Flash Lamp Annealing. Nanomaterials (2020) 10: 648.

[126] Krishna Nichenametla C., Calvo J., Riedel S., Gerlich L., Hindenberg M., Novikov S., et al. Doping Effects in CMOS ‐ compatible CoSi Thin Films for Thermoelectric and Sensor Applications. Zeitschrift für anorganische und allgemeine Chemie (2020) 646: 1231-1237.

[127] Calamba K. M., Salamania J., Jõesaar M. P. J., Johnson L. J. S., Boyd R., Pierson J. F., et al. Effect of nitrogen vacancies on the growth, dislocation structure, and decomposition of single crystal epitaxial $(Ti_{1-x}Al_x)N_y$ thin films. Acta Materialia (2021) 203: 116509.

[128] Bain M. F., Deo N., Armstrong B. M., Gamble H. S. Effect of deposition temperature on the formation of $CoSi_2$ through the rapid thermal annealing of CVD cobalt. Microelectronic Engineering (2004) 76: 336-342.

[129] Luo Q., Yang S. Uncertainty of the X-ray Diffraction (XRD) $sin^2 \psi$ Technique in Measuring Residual Stresses of Physical Vapor Deposition (PVD) Hard Coatings. Coatings (2017) 7: 128.

[130] Davidse P. D., Maissel L.I. Dielectric Thin Films through rf Sputtering. Journal of Applied Physics (1966) 37: 574.

[131] Gudmundsson J. T., Brenning N., Lundin D., Helmersson U. High power impulse magnetron sputtering discharge. Journal of Vacuum Science & Technology A: Vacuum, Surfaces, and Films (2012) 30: 030801.

[132] Son H. H., Seo G. H., Jeong U., Shin D. Y., Kim S. J. Capillary wicking effect of a Cr-sputtered superhydrophilic surface on enhancement of pool boiling critical heat flux. International Journal of Heat and Mass Transfer (2017) 113: 115-128.

[133] Li Z., Xie Y., Yuan Y., Wang M., Xu C, Hübner R., et al. B20-type FeGe on Ge(1 0 0) prepared by pulsed laser melting. Journal of Magnetism and Magnetic Materials (2021) 532: 167981.

[134] Rebohle L., Neubert M., Schumann T., Skorupa W. Determination of the thermal cycle during flash lamp annealing without a direct temperature measurement. International Journal of Heat and Mass Transfer (2018) 126: 1-8.

[135] Zhang L., Ivey D. G. Low temperature reactions of thin layers of Mn with Si. Journal of materials research (1991) 6: 1518-1531.

[136] Connolly J. R. Introduction to X-ray Powder Diffraction. Spring 2005.

[137] Buchner M., Höfler K., Henne B., Ney V., Ney A. Tutorial: Basic principles, limits of detection, and pitfalls of highly sensitive SQUID magnetometry for nanomagnetism and spintronics. Journal of Applied Physics (2018) 124: 161101.

[138] Dzyaloshinskii I. E. Thermodynamic Theory of 'Weak" Ferromagnetism in Antiferromagnetic Substances. Sov Phys JETP (1957) 5: 1259.

[139] Moriya T. Anisotropic Superexchange Interaction and Weak Ferromagnetism. Physical Review (1960) 120: 91-98.

[140] Raju M., Yagil A., Soumyanarayanan A,, Tan A. K. C., Almoalem A., Ma F., et al. The evolution of skyrmions in Ir/Fe/Co/Pt multilayers and their topological Hall signature. Nature Communications (2019) 10: 696.

[141] Stolt M. J., Schneider S., Mathur N., Shearer M. J., Rellinghaus B., Nielsch K., et al. Electrical Detection and Magnetic Imaging of Stabilized Magnetic Skyrmions in $Fe_{1-x}Co_xGe$ (x < 0.1) Microplates. Advanced Functional Materials (2019) 29: 1805418.

[142] Yang M., Li Q., Chopdekar R. V., Dhall R., Turner J., et al. Creation of skyrmions in van der Waals ferromagnet. Science Advances (2020) 6: 36 eabb5157.

[143] Li Z., Xie Y., Yuan Y., Ji Y., Begeza V., Cao L., et al. Phase Selection in Mn–Si Alloys by Fast Solid–State Reaction with Enhanced Skyrmion Stability. Advanced Functional Materials (2021) 31: 2009723.

[144] Luo Y., Lin S-Z., Leroux M., Wakeham N., Fobes D. M., Bauer E. D., et al. Skyrmion lattice creep at ultra-low current densities. Communications Materials (2020) 1: 83.

[145] Zhang S. L., Chalasani R., Baker A. A., Steinke N. J., Figueroa A. I., Kohn A., et al. Engineering helimagnetism in MnSi thin films. AIP Advances (2016) 6: 015217.

[146] López-López J., Gomez-Perez J. M., Álvarez A., Babu Vasili H., Komarek A. C., Hueso L. E., et al. Spin fluctuations, geometrical size effects, and zero-field topological order in textured MnSi thin films. Physical Review B (2019) 99: 14427.

[147] Zhou R., Feng M., Wang J., Sun Q., Liu J., Zhang S., et al. InGaN-Based Lasers with an Inverted Ridge Waveguide Heterogeneously Integrated on Si(100). ACS Photonics (2020) 7: 2636-2642.

[148] Hortamani M., Sandratskii L., Kratzer P., Mertig I., Scheffler M. Exchange interactions and critical temperature of bulk and thin films of MnSi: A density functional theory study. Physical Review B (2008) 78: 104402.

[149] Kim G., Kim H-S., Lee H. S., Kim J., Lee K. H., Roh J. W., et al. Synchronized enhancement of thermoelectric properties of higher manganese silicide by introducing Fe and Co nanoparticles. Nano Energy (2020) 72: 104698.

[150] Hou Q. R., Zhao W., Chen Y. B., He Y. J. Preparation of n-type nano-scale $MnSi_{1.7}$ films by addition of iron. Materials Chemistry and Physics (2010) 121: 103-108.

[151] Lian Y. C., Chen L. J. Localized epitaxial growth of $MnSi_{1.7}$ on silicon. Applied Physics Letters (1986) 48: 359-61.

[152] Teicherta S., Schwendler S., Sarkar D. K., Mogilatenko A., Falke M., Beddies G., et al. Growth of $MnSi_{1.7}$ on Si(0 0 1) by MBE. Journal of Crystal Growth (2001) 227: 882-887.

[153] Kahwaji S., Gordon R. A., Crozier E. D., Monchesky T. L. Local structure and magnetic properties of B2- and B20-like ultrathin Mn films grown on Si(001). Physical Review B (2012) 85: 014405.

[154] Wu H., Hortamani M., Kratzer P., Scheffler M. First-Principles Study of Ferromagnetism in Epitaxial Si-Mn Thin Films on Si(001). Physical Review Letters (2004) 92: 237202.

[155] Xie Y., Yuan Y., Wang M., Xu C., Hübner R., Grenzer J, et al. Epitaxial Mn_5Ge_3 (100) layer on Ge (100) substrates obtained by flash lamp annealing. Applied Physics Letters (2018) 113: 222401.

[156] Bechler S., Kern M., Funk H. S., Colston G., Fischer I. A., Weißhaupt D., et al. Formation of Mn_5Ge_3 by thermal annealing of evaporated Mn on doped Ge on Si(111). Semiconductor Science and Technology (2018) 33: 095008.

[157] Zi J., Büscher H., Falter C., Ludwig W., Zhang K., Xie X. Raman shifts in Si nanocrystals. Applied Physics Letters (1996) 69: 200-202.

[158] Tite T., Shu G. J., Chou F. C., Chang Y. M. Structural and thermal properties of MnSi single crystal. Applied Physics Letters (2010) 97: 031909.

[159] Eiter H. M., Jaschke P., Hackl R., Bauer A., Gangl M., Pfleiderer C. Raman study of the temperature and magnetic-field dependence of the electronic and lattice properties of MnSi. Physical Review B (2014) 90: 024411.

[160] Ferri F. A., Pereira-da-Silva M. A., Zanatta A. R. Development of the $MnSi_{1.7}$ phase in Mn-containing Si films. Materials Chemistry and Physics (2011) 129: 148-153.

[161] Bürger D., Baunack S., Thomas J., Oswald S., Wendrock H., Rebohle L., et al. Evidence for self-organized formation of logarithmic spirals during explosive crystallization of amorphous Ge:Mn layers. Journal of Applied Physics (2017) 121: 184901.

[162] Dhital C., Khan M. A., Saghayezhian M., Phelan W. A., Young D. P., Jin R. Y., et al. Effect of negative chemical pressure on the prototypical itinerant magnet MnSi. Physical Review B (2017) 95: 024407.

[163] Kazuo K., Muneyuki D. magnetization and magnetoresistance of MnSi1. Journal of the Physical Society of Japan (1982) 52: 2433.

[164] Neubauer A., Pfleiderer C., Ritz R., Niklowitz P. G., Böni P. Hall effect and magnetoresistance in MnSi. Physica B: Condensed Matter (2009) 404: 3163-3166.

[165] Samatham S. S., Ganesan V. Critical behavior, universal magnetocaloric, and magnetoresistance scaling of MnSi. Physical Review B (2017) 95: 115118.

[166] Das B., Balasubramanian B., Skomski R., Mukherjee P., Valloppilly S. R., Hadjipanayis G. C., et al. Effect of size confinement on skyrmionic properties of MnSi nanomagnets. Nanoscale (2018) 10: 9504-9508.

[167] Bogdanov A. N., Yablonskii D. A. Thermodynamically stable "vortices" in magnetically ordered crystals. The mixed state of magnets. Sov Phys JETP (1989) 68: 178-182.

[168] Rößler U. K., Bogdanov A. N., Pfleiderer C. Spontaneous skyrmion ground states in magnetic metals. Nature (2006) 442: 797-801.

[169] Jonietz F., Mühlbauer S., Pfleiderer C., Neubauer A., Münzer W., Bauer A., et al. Spin transfer torques in MnSi at ultralow current densities. Science (2009) 330: 6011 1648-1651.

[170] Neubauer A., Pfleiderer C., Binz B., Rosch A., Ritz R., Niklowitz P. G., et al. Topological Hall Effect in the A Phase of MnSi. Physical Review Letters (2009) 102: 186602.

[171] Higashi S., Ikedo Y., Kocán P., Tochihara H. Epitaxially grown flat MnSi ultrathin film on Si(111). Applied Physics Letters (2008) 93: 013104.

[172] Pierre Villars (Chief Editor), PAULING FILE in: Inorganic Solid Phases, Springer science Materials (online database), Springer, Heidelberg (ed.).

[173] Li Z., Dong J-F., Sun F-H., Pan Y., Wang S-F., et al. MnS Incorporation into Higher Manganese Silicide Yields a Green Thermoelectric Composite with High Performance/Price Ratio. Advanced Science (2018) 5: 1800626.

[174] Kissinger H. E., Reaction kinetics in differential thermal analysis. Anal. Chem. (1957) 29: 1702–1706.

[175] Kosiba K., Pauly S. Inductive flash-annealing of bulk metallic glasses. Scientific Reports (2015) 7: 2151.

[176] Monnier X., Cangialosi D., Ruta B., Busch R., Gallino S. Vitrification decoupling from α-relaxation in a metallic glass. Science Advances (2020) 6: eaay1454.

[177] Sohn S., Xie Y., Jung Y., Schroers J., Cha J. J. Tailoring crystallization phases in metallic glass nanorods via nucleus starvation. Nature Communications (2017) 8: 1980.

[178] Salvo M., Smeacetto F., D'Isanto F., Viola G., Demitri P., Gucci F., et al. Glass-ceramic oxidation protection of higher manganese silicide thermoelectrics. Journal of the European Ceramic Society (2019) 39: 66-71.

[179] Hou Z., Ren W., Ding B., Xu G., Wang Y., Yang B., et al. Observation of Various and Spontaneous Magnetic Skyrmionic Bubbles at Room Temperature in a Frustrated Kagome Magnet with Uniaxial Magnetic Anisotropy. Advanced Materials (2017) 29: 1701144.

[180] Okuyama D., Bleuel M., White J. S., Ye Q., Krzywon J., Nagy G., et al. Deformation of the moving magnetic skyrmion lattice in MnSi under electric current flow. Communications Physics (2019) 2: 79.

[181] Zhou S., Potzger K., Zhang G., Mücklich A., Eichhorn F., Schell N., et al. Structural and magnetic properties of Mn-implanted Si. Physical Review B (2007) 75: 085203.

[182] Gluschenkov O., Jagannathan H. Laser Annealing in CMOS Manufacturing. ECS Trans (2018) 85: 11.

[183] Taniguchi T., Ishibe T., Naruse N., Mera Y., Alam M. M., Sawano K., et al. High Thermoelectric Power Factor Realization in Si-Rich SiGe/Si Superlattices by Super-Controlled Interfaces. ACS Applied Materials & Interfaces (2020) 12: 25428-25434.

[184] Zantye P. B., Kumar A., Sikder A. K. Chemical mechanical planarization for microelectronics applications. Materials Science and Engineering: R: Reports (2004) 45: 89-220.

[185] Bak P., Jensen M. H. Theory of helical magnetic structures and phase transitions in MnSi and FeGe. Journal of Physics C: Solid State Physics (1980) 13: L881.

[186] Pfleiderer C., Thessieu C., Stepanov A. N., Lapertot G., Couach M., Flouquet J. Field dependence of the magnetic quantum phase transition in MnSi. Physica B (1997) 230: 576-579.

[187] Pfleiderer C., McMullan G. J., Julian S. R., Lonzarich G. G. Magnetic quantum phase transition in MnSi under hydrostatic pressure. Physical Review B (1997) 55: 8330.

[188] Ritz R., Halder M., Wagner M., Franz C., Bauer A., Pfleiderer C. Formation of a topological non-Fermi liquid in MnSi. Nature (2013) 497: 231-234.

[189] Potapova N., Dyadkin V., Moskvin E., Eckerlebe H., Menzel D., Grigoriev S. Magnetic ordering in bulk MnSi crystals with chemically induced negative pressure. Physical Review B (2012) 86: 060406(R).

[190] Meyer T. L., Jiang .L, Park S., Egami T., Lee H. N. Strain-relaxation and critical thickness of epitaxial La1.85Sr0.15CuO4 films. APL Materials (2015) 3: 126102.

[191] Peng L. S. J., Xi X. X., Moeckly B. H., Alpay S. P. Strain relaxation during in situ growth of SrTiO3 thin films. Applied Physics Letters (2003) 83: 4592-4594.

[192] Thomas O., Shen Q., Schieffer P., Tournerie N., Lépine B. Interplay between Anisotropic Strain Relaxation and Uniaxial Interface Magnetic Anisotropy in Epitaxial Fe Films on (001) GaAs. Physical Review Letters (2003) 90: 017205.

[193] Yu G., Jenkins A., Ma X., Razavi S. A., He C., Yin G., et al. Room-Temperature Skyrmions in an Antiferromagnet-Based Heterostructure. Nano Letters (2018) 18: 980-986.

[194] Khan S., Zollitsch C. W., Arroo D. M., Cheng H., Verzhbitskiy I., Sud A., et al. Spin dynamics study in layered van der Waals single-crystal $Cr_2Ge_2Te_6$. Physical Review B (2019) 100: 134437.

[195] Zhang X., Wang B., Guo Y., Zhang Y., Chen Y., Wang J. High Curie temperature and intrinsic ferromagnetic half-metallicity in two-dimensional Cr_3X_4 (X = S, Se, Te) nanosheets. Nanoscale Horizons (2019) 4: 859-866.

[196] Markovich V., Fita I., Wisniewski A., Puzniak R., Martin C., Jung G., et al. Phase transitions and magnetic properties of $LuFe_2O_4$ under pressure. Physical Review B (2017) 96: 054416.

[197] Balasubramanian B., Manchanda P., Pahari R., Chen Z., Zhang W., Valloppilly S. R., et al. Chiral Magnetism and High-Temperature Skyrmions in B20-Ordered Co-Si. Physical Review Letters (2020) 124, 057201.

[198] Yin L., Wang C., Shen Q., Zhang L. Strain-induced Curie temperature variation in $La_{0.9}Sr_{0.1}MnO_3$thin films. RSC Advances (2016) 6: 96093-96102.

[199] Lu W., Song W., Yang P., Ding J., Chow G. M., Chen J. Strain Engineering of Octahedral Rotations and Physical Properties of $SrRuO_3$ Films. Scientific Reports (2015) 5: 10245.

[200] Théry V., Boulle A., Crunteanu A., Orlianges J. C. Combined strain and composition-induced effects in the metal-insulator transition of epitaxial VO_2 films. Applied Physics Letters (2017) 111: 251902.

[201] Zhang H., Gu L., Zhang L., Zheng S., Wan H., Sun J., et al. Removal of aqueous Pb(II) by adsorption on Al_2O_3 -pillared layered MnO_2. Applied Surface Science (2017) 406: 330-338.

[202] Welzel U., Ligot J., Lamparter P., Vermeulen A. C., Mittemeijer E. J. Stress analysis of polycrystalline thin films and surface regions by X-ray diffraction. Journal of Applied Crystallography (2005) 38: 1-29.

[203] Yabuuchi S., Kageshima H., Ono Y., Nagase M., Fujiwara A., Ohta E. Origin of ferromagnetism of $MnSi_{1.7}$ nanoparticles in Si: First-principles calculations. Physical Review B (2008) 78: 045307.

[204] Men'shov V. N., Tugushev V. V., Caprara S., Chulkov E. V. High-temperature ferromagnetism in Si:Mn alloys. Physical Review B (2011) 83: 035201.

[205] Rylkov V. V., Nikolaev S. N., Chernoglazov K. Y., Aronzon B. A., Maslakov K. I., Tugushev V. V., et al. High-temperature ferromagnetism in $Si_{1-x}Mn_x$ (x $\approx$ 0.5) nonstoichiometric alloys. JETP Letters (2012) 96: 255-262.

[206] Tuomisto F., Makkonen I. Defect identification in semiconductors with positron annihilation: Experiment and theory. Reviews of Modern Physics (2013) 85: 1583-1631.